Terror by Error?

The Covid Chronicles

William Sargent

Cover by Jill Buchanan

"Yes but while we are talking, scientists are are working in hidden secret laboratories trying to develop, as best they can, diseases which the other man can't cure."

Table of Contents

Acknowledgements

Many people over dozens of years have helped me understand the murky, intersecting worlds of medical research and biological warfare. Both are places where mistakes are made that sometimes lead to medical breakthroughs, sometimes to deadly pandemics.

When I was writing about the push to develop vaccines for troops entering Iraq, Tom Monath of Acambis introduced me to his world where grueling fieldwork led him to form a small start-up company that beat out much larger firms to win a hefty contract to develop a new vaccine for smallpox.

Ken Alibek told me about his journey from being an idealistic young medical student to being the head of Biopreparat, the Soviet Union's dual use facility that produced hundreds of tons of biological agents that were loaded onto missiles targeting for cities in the United States.

After defecting, Ken became a successful consultant and owner of a multi-million dollar biotech company in Virginia. The fact that US intelligence never knew the true nature of Biopreparat gave me pause about the origins of Covid-19 as well.

Sue little, the owner of the inestimable Newburyport bookstore Jabberwocky, and Don Paquin told me of their decades long bouts with tick-borne diseases. Carl Soderland who gave me insights about treating these controversial new illnesses.

I would particularly like to thank the late John Buttman from the Ike Williams agency who believed in this book from the beginning. I was especially looking forward to working with him when he was struck down by a heart attack in the early days of the pandemic.

I would also like to thank Becky Coburn who has now stewarded six of my manuscripts through the entire publication process, and my friend and inestimable editor Richard Lodge at the Newburyport Daily where many of these chapters were first published.

Finally I would like to thank the good people at the Quebec Labrador Foundation, The Sounds Conservancy, Andy Griffith, and Plum Island Outdoors for providing grants to help fund research for this book.

DEDICATION

To Luna.

I hope you will live in a better world because of how we respond to Covid-19.

INTRODUCTION
The Diamond Princess - Yokohama, Japan
February 2020

Jerri Serratti-Goldman was elated to finally be going on a cruise through the Orient with her husband Carl. He had given her tickets for passage on the *Diamond Princess* as a Christmas present and they had departed on their dream vacation on January 17.

But two days before the end of the 16-day cruise the Captain announced that a passenger who had already departed from the ship had been diagnosed with a coronavirus so they would have to return to Yokohama and go into quarantine for 12 more days.

They had a sunny starboard cabin and could look through their portholes and see ambulances rushing passengers to the Yokohama Hospital. But a steady stream of crew members in masks and gloves brought them trays full of delicious food, so they decided to just make the best of it.

The Goldmans were fortunate. Carl had booked an expensive upper deck cabin so they could sit on their balcony and walk the deck three times a day.

Passengers in the lower decks had no windows and their cabins were so cramped they had to lie in their bunks or sit on a straight-backed chair breathing infectious, recycled air.

The U. S. State department finally rescued the couple, flying them out of Japan on a chartered plane, but two hours into the flight Carl awoke. "I had a high fever after napping for a little bit."

He was flown on to Omaha with 12 other infected passengers to be quarantined in a bio containment unit where his only contact with the outside world were staff who could just wave at him through a thick window and doctors in Hazmat suits who gave him no medications but checked his vital signs and gave him Gatorade three times a day.

The new coronavirus had apparently emerged on December 1 2019. At first it looked like the novel virus, dubbed Covid-19, had originated in the Wuhan seafood market where people bought snakes and Pangolins that had been eating infected bats.

The disease had spread rapidly and by mid-February Wuhan was in strict lockdown and the Covid-19's annual rate of infection was through the roof.

It was this annual rate of infection that drew me into investigating the infectious new pathogen because it was so similar to the infection rate of the tick borne diseases that had been ravaging the East Coast.

But there was a big difference between the two epidemics. Covid-19 had evolved a mutation so it could be spread from human to human, perhaps from the genome of one of its human victims, perhaps from the genome of one of its bat, snake or pangolin vectors. However tick borne diseases had never evolved the ability to be transmitted from human to human.

While I was doing research on the outbreak I came across a curious online comment that said that the virus had probably escaped from a Level 4 Bio Safety facility in Wuhan China and that it was the only such facility in China. Now, this was interesting. I had never heard of the facility and decided to investigate further.

It turned out that Israeli intelligence and several biological weapons experts believed that the Wuhan lab is a dual-use facility like the Plum Island Animal Research Center was before 1969.

That was when Plum Island housed one lab operated by the U.S. Health Service to develop vaccines and another lab operated by the U.S. Army to develop biological weapons.

Scientists at the Wuhan facility study some of the world's most dangerous pathogens and they have been implicated in stealing coronaviruses from a Winnipeg lab to add to their stock of vaccines…or weapons.

Is it possible that Wuhan scientists used the new CRISPR gene splicing technology to tweak their coronaviruses so they could be transmitted from human to human?

Could it be that so-called patient zero had not been infected by a bat from the Wuhan Seafood Market, but was a technician who had been accidentally infected at the Wuhan facility and had then spread the virus to the surrounding human population?

There was also a rumor flying around pharmaceutical circles that someone in the facility had sold research animals to the Wuhan market after the scientists were done with them. And this wasn't just a janitor making a few extra Yuan on the side. A technician in the Wuhan Center for Virology claimed the head of the lab had sold over a million dollars worth of used research animals to food vendors.

If these scenarios prove to be true, what happened at the Wuhan facility could be similar to what happened when disease-carrying ticks escaped from the Plum Island Animal Disease Center and were spread up and down the East Coast by shorebirds.

Both instances would prove that General Eisenhower's prophetic observation was correct. He said that the problem with biological weapons was that they could blow back onto your own troops, and spread to the civilian population. What he did not realize was that this could happen on a global scale.

On February 26, the stock market faltered after the CDC announced that "It was not a matter of if, but when" the coronavirus would be transmitted to the United States killing off perhaps 2 percent of its population, and it was "inevitable" that this would happen.

President Trump was furious with the announcement, which had been released, probably purposefully, when he was on a state visit to India. He wondered out loud whether Alex Azar, the head of Health and Human Services should resign and held a quickly pulled together press conference on his return.

It was a classic Trumpian performance. To make his point that he was upset with the doctors and scientists at the CDC and HHS he put his most subservient "yes man" Michael Pence in charge of dealing with the virus and stated that his bans on travel from China and Japan was the reason that only 15 Americans, including Carl Goldman, had come down with the virus. There were actually 60 other cases he neglected to mention.

But the sleep deprived President forged on, "I can't tell you if it's going to get better or get worse. It is probably going to get better and go away. I don't think it is inevitable that it will spread through the United States. But if it does we are ready willing and able to respond."

"It's a little like the regular flu that we have flu shots for. And we will essentially have flu shot for this in a fairly quick manner."

The statements were in direct contradiction to the CDC's announcement and shortly after the president finished, Anthony Fauci, Head of the National Institute for Allergy and Infectious Diseases, corrected the president by telling journalists that despite receiving a candidate vaccine from the Moderna Inc., it would be a year to a year and a half before a vaccine could go through trials and be available for widespread use.

In the meantime schools, businesses and hospitals should prepare plans to respond to a major medical crisis.

It was clear that what Trump was really concerned about was that the stock market crash would affect his bid for reelection. He even blamed the crash on the Democratic debate, which was curious, since it had been held after the market had started its decline.

But Trump had reason to be concerned about his reelection chances. This would be the first time his leadership would be tested by a national crisis like Katrina or 9/11.

He would learn how risky it is to make political calculations based on the behavior of such an erratic virus. And, by the end of February 2020, Covid-19 was an extremely erratic and successful pathogen, precisely because it was so communicable but not very deadly. This meant that it could spread rapidly all over the globe without killing off its human hosts and thus itself.

But that could change. Covid-19 could evolve to become even more communicable and even less deadly... or evolve to become less communicable but much more deadly. And we probably wouldn't know that until after the fall elections.

If Trump wanted to learn about the risks of making political decisions based on virus behavior, he needed to look no further than his fellow Republican President Gerald Ford. Ford had been in danger of losing his bid for reelection because he had pardoned Nixon and was casting around for something to do that would be universally favorable.

He decided to inoculate "every man, woman and child" against Swine flu that was expected to be severe in 1976. But the flu never materialized and the vaccines against it were contaminated because they had been rushed into production without adequate testing. Thousands of people ended up contracting severe neurological problems and President Ford lost the election to Jimmy Carter who had been given little chance of winning before the great Swine flu fiasco.

So to truly understand how Covid-19 originated and spread we will have to go back hundreds of years to the beginning of the United States and thousands of years to China's Celestial Kingdom.

Chapter i
The Death of a Nation, Patuxet
1617

The sun rose slowly out of the Atlantic Ocean. It glistened off the surface of Patuxet Bay and highlighted the tidy fields and open forests of the Wampanoag lands. It was 1617; the Native American tribe was at its peak of influence. People traveled freely from village to village, trade flourished and the supreme sachem settled any disputes in his far-flung domain.[1]

But there were no disputes this morning. Attaquin, a young warrior, stepped out of his long house to greet the newborn day. He was a tall quiet man with clear-bronzed skin and a reserved manner. At his feet lay a pile of clam and lobster shells, the remains of last night's feast. He swept the shells into a reed basket and dumped them at the edge of the village before returning to recline on his pile of deerskin blankets. He enjoyed sitting in the early morning sun watching the village as it slowly awoke.

All that could be heard was the quiet gabbling of a skein of geese as they flew purposefully through the cobalt clear sky. The harvest had been good this year. The fields were covered with the stubble of last summer's corn. Beyond the fields lay the open woods. Every year the warriors burned the underbrush to keep the forest open so they could hunt deer and turkey in the tamed landscape.

Below the village lay the quiet waters of Patuxet Bay.[2] They lapped against the distant shore and shimmered in the early morning sun. It was the bay that provided the Patuxet with the crabs, fish and shellfish that made up the bulk of their nutritious diet. Hundreds of similar villages hugged this shore that would soon be called Cape Cod and Southern New England, now it was simply the loose confederation of villages known as the Wampanoag Nation.

Things were good in the nation mused Attaquin. The land and water provided the Wampanoag with all the food they could eat. Ashumet ruled the village well; the supreme sachem had kept the nation out of war for as long as anyone could remember.[3]

But now there was a commotion at the sachem's lodge. A runner from a neighboring village arrived with welcome news. A pod of small whales had come ashore on the Great Beach of the Nausets.

This was the moment Attaquin had been waiting for. For many months he had carefully hammered and chipped his long stone knives. Now they were thirteen inches long with beautiful fluted edges. They looked like fine ceremonial objects, but Attaquin had other plans for his finely wrought tools.[4]

Attaquin and his brother were recognized as the best sword fishermen in the village. Several years before they had spent months alternately chopping and burning the base of a thick pine tree they had spotted towering over the surrounding canopy. After it had thundered to the ground they spent several more weeks laboriously hacking the interior of the log with adzes and burning it with small fires. When they finally finished their dugout, it was long, wide and sturdy enough to take the brothers far out into the Atlantic Ocean in search of the giant fish that slumbered on the surface.

Attaquin and his brother had learned to work as a team. Attaquin would stand in the bow giving hand signals so Quitsa could quietly paddle up behind the swordfish before it knew the brothers were there. The men would then hold their breath as Attaquin quietly slipped his spear into his atl-atl, before thrusting it deep into the back of the somnolent giant. Bladders filled with air would eventually slow the fish enough so that Attaquin could deliver the coup de grace.

The entire village would assemble when the brothers returned with their catch. Attaquin would clean the fish and Quitsa would distribute pieces to every member of the band.

He did this skillfully; garnering praise, and political obligations. The sachem thanked the brothers, the elders praised them, and young women vied to catch their attention.

Now this new discovery of whales would give Quitsa and Attaquin another chance to provision the village and flaunt their prowess before the neighboring Nausets. With rising excitement the brothers loaded their dugout with nets for catching fish, atl-atls for throwing their harpoons, and adzes in case they had to build another dugout. It was good to be prepared for any eventuality. Finally Attaquin wrapped his long knives in deerskin, packed them carefully in the bow of the dugout, and pushed it out into the bay.

The brothers were soon joined by three more dugouts that rounded the barrier beach and paddled purposefully down the coast. Their destination was the inner arm of Cape Cod, a full day's paddle away. Miles before they reached the great beach Attaquin could see the dark forms of whales stranded on the sandflats. Each carcass was surrounded by half a dozen men hacking at the whales' tough skin. As they approached, Attaquin identified himself as a Patuxet and announced that he had come for his village's share of the whales.

Pashtum, Attaquin's old friend from the Nauset tribe, recognized him instantly. He chided Attaquin for being so formal and arriving so late. Only the scrawniest of whales were left to butcher and the Patuxets would surely lose their loincloths in the ball game that was to follow.

After much bantering, Attaquin and the Patuxets paddled to the section of beach reserved for them by treaty. Soon they were busily stripping great slabs of muscle-rich red meat off the whales' bloody carcasses. Warriors from other villages wandered over to admire Attaquin's long stone knives that cut so deeply through the thick layers of blubber. He could dress out two whales to their one.

Pashtum offered to buy one of Attaquin's knives. They were even better than the metal knife that Pashtum had bought from a white man.

White men were now appearing along these shores with some frequency. Englishmen and Spaniards came to fish in the summer months before sailing back to Europe. The villagers viewed the strangers with a mixture of curiosity and suspicion. Everyone agreed that the white man's food was atrocious. The villagers ate it to be polite but it made them tired and weak. One villager had just come down with red bumps from eating the white man's food. They joked that giving the white men good food was probably a big mistake. Having once tasted good Wampanoag food the white men would never want to leave.[5]

But soon the butchering was over and the villages separated into two teams. The warriors joked and jostled as they battled with sticks and their bodies to drive a small ball up and down several miles of beach. The game continued for four days. After it was over the warriors hung their clothes, spears and wampum on a makeshift arbor and rolled pieces of antler to gamble on each other's wares. The point of the contest was to bet, laugh and swear loudly at the outcome. On the last night the villages danced and Attaquin slept with Pashtum's playful sister Nanatatuck.[6]

The following morning Attaquin and the vanquished Patuxets climbed into their dugouts to paddle back to their village. Nanatatuck and her friends lined up on the bank to laugh and admire their new friends as they paddled away naked. The Patuxet vowed that next year they would return and it would be the Nausets who would lose their loincloths.

Perhaps it was from lack of sleep, perhaps it was from missing Nanatatuck, perhaps it was from paddling in the cold but Attaquin was strangely quiet on the way home. By the time they reached Patuxet Bay he knew that something was seriously wrong. He had a stabbing pain in base of his spine and his head was starting to throb.

But there was work to be done. After unloading, the brothers repacked the dugout with spears and Attaquin's new flensing knives, covered it with deerskins, and buried it under soil and leaves. The dugout would be waiting for them, packed and ready when the swordfish returned in the spring.

By nightfall Attaquin was exhausted. He vomited before crawling into the back of the long house to sleep. But sleep did not come easily. Attaquin had frightening dreams of sick and dying whales. Had the Patuxet not thanked the whales for honoring the Indians with their meat? Inside his body, tiny particles of virus were starting to shock Attaquin's immune system and disturb his dreams.[7]

But the following morning, people wanted to hear about Nauset and the whales. Attaquin propped himself up on deerskins and regaled his visitors with stories. One by one they leaned close to hear, one by one they entered the miasma of contagion that hung about Attaquin like an invisible cloud. As he spoke, tiny bits of virus leaked out of the back of Attaquin's throat and spewed out on droplets of his saliva. One of the virions landed on the mucus linings of Quitsa's throat and came alive. It bore through his cell wall and started to replicate, amplifying throughout his body. Quitsa was the second wave of the viral attack. He would not feel ill for a fortnight.

The same could not be said for Attaquin. By the end of the day his skin was covered with thousands of tiny red dots and he was starting to wretch and gag up copious quantities of foul liquid vomit. Two days later Attaquin awoke screaming in pain. Scores of dark patches had merged so his skin looked like it had been charred in a fire. Whenever he moved, hideous patches of skin would peel off and stick to his deerskin blanket.

Attaquin now wore what diagnosticians call the "worried look of smallpox." The pain was excruciating and the village could not escape the piercing sounds of his moans and screams. Even the smoke of the medicine man's fires could not dispel the putrid odor of Attaquin' s rotting skin.

The smallpox had so shocked his immune system that Attaquin could no longer make pus and the oozing pustules of Variola major. This was hemorrhagic smallpox, black smallpox, the disease where blood the color of coal seeps uncoagulated beneath the skin and oozes from the mouth and eyelids.

Attaquin was paralyzed with fright. He stared at his visitors with wild, bloodshot eyes. He felt as though his insides were coming apart - and in a sense they were. The disease was destroying his skin both inside and out. It had already liquefied the linings of his throat, stomach, intestines and rectum.

A spasm coursed down the length of Attaquin's body and he lurched forward. Blood speckled with bits of liquefied intestine gushed out of his anus. The stench was terrible. Quitsa and his mother hid their horror as they wrapped Attaquin in his stained deerskin and carried him outside to a nearby sweat lodge. Attaquin was still conscious and in excruciating pain, but the fever was too much. He threw off the deerskin and staggered into the icy cold waters of the Patuxet River. Hopefully the waters provided Attaquin with some small relief before they triggered the heart attack that ended his life forever.

Fourteen days later Quitsa and nine other villagers had also perished, but not before passing on the deadly virus. Wave after wave of villagers died. Soon there was no one left to feed the ill or bury the dead. Corpses piled up inside long houses and outside in the snow where the ill had fled to escape their fevers.

By the end of winter it was mostly over. Attaquin's entire village had been swept clean by the ravaging illness. All that could be seen were feral dogs searching through empty lodges, gnawing the corpses of their former owners. Up and down the coast it was the same story. Dead and dying villages, ninety percent of the people gone, fields lying fallow, forests encroaching. This once thriving nation was no more. It could cover up its weaknesses and welcome the Pilgrims but it would never regain what it had lost to the Europeans and their smallpox.[8]

The Pilgrims only became aware of the extent of tragedy when they visited the Narragansett tribe several years after the demise of the Patuxet. They noticed that they could walk freely among the dead and dying Indians without themselves being infected. As Governor Bradford wrote:

"But those of the English house, though they were afraid of the infection, yet seeing their woeful and sad condition and hearing their pitiful cries and lamentations, had compassion of them and daily fetched them wood and water, made their fires, got them victuals whilst they lived; and buried them when they died. For very few of them escaped ... But by the marvelous goodness and providence of God, not one of the English was so much as sick or in the least measure tainted with this disease, though they did daily do these offices for them for many weeks together".[9]

The Wampanoag had fallen victim to what modern epidemiologists call a "virgin soil" epidemic. The Wampanoags had never been exposed to smallpox so their immune systems had never been primed to recognize or fight the contagion. The Pilgrims, on the other hand, had come from the cities of England and Europe where smallpox ran rampant and where at least some people had developed a natural immunity to the disease.

One of the witnesses to this tragedy was Squanto, that supposed paragon of Native American hospitality. But there was a darker side to this patron saint of Thanksgiving. As a young man, Squanto had been captured by explorers and shipped to Spain to be sold into slavery. No one knows precisely what happened to Squanto during his six years wandering through Europe. We do know, however, that he escaped the fate of his village, Patuxet, and we can surmise that he had probably been exposed to a milder strain of smallpox than what befell his brethren.

When he returned to Patuxet, Squanto found his village had been destroyed as had most of the rest of the Wampanoag nation. Southern New England was a desolation of half-empty villages and fields returning to forest. With his nation crippled and his village destroyed Squanto decided to side with the Pilgrims. It was not a bad choice. He had no work and the Pilgrims had no translator. But Squanto, like many who would follow, was not above using the situation to his own advantage.

In dealing with his fellow Indians, Squanto was known to threaten that the English kept smallpox buried in the ground and could call it out at will. The English didn't entirely discredit the idea. When an Indian asked about a suspicious bag of powder that he had seen the Pilgrims bury, they assured him it was gunpowder not smallpox, "But our God does have powere over the plague and can use it against our enemies". [10]

It was not the first time that someone claimed that God was on their side, and it would not be the last time that an entire nation would be wiped off the face of the planet by a deadly germ.

CHAPTER 2
Washington's Gamble
Variolation 1777

During the American Revolution George Washington faced a dilemma. He knew that most of the British troops had gained immunity by being exposed to smallpox in England while his American born troops were as susceptible to the disease as the Wampanoags had been prior to the settlement of the Plymouth Colony. He also knew that the British were not above using the situation to their advantage. Lord Amherst had ordered that the Patriots' Indian allies be given smallpox contaminated blankets prior to the Battle for Quebec and a British author had written that the redcoats should dip arrows in smallpox and "twang" them at the American troops.[1]

However, Washington's methods for fighting smallpox were limited. When he had General Howe bottled up in Boston in 1775; he dared not attack the British troops because smallpox was running rampant through the city. Washington feared that if he attacked, his troops would come down with smallpox and his entire army would be decimated. His only recourse was to institute a strict quarantine. He prohibited anyone from the city from visiting his troops and he ordered all army mail to be dipped in vinegar before delivery. Civilians with symptoms of smallpox were quarantined in a hospital in Brookline while military cases were sent to a separate hospital in Cambridge near Fresh Pond, the army's primary source of water. Even citizens deemed smallpox free had to walk through smoke and be thoroughly cleansed before they could receive written confirmation that they were free of the illness.[2]

At the height of the pestilence, spies reported that thirty to forty people a day were dying from smallpox in Boston, but that Lord Howe had forbidden the tolling of bells during funerals so Washington could not learn the true extent of the epidemic.[3]

A few months before he gave up the city Lord Howe instituted an invidious campaign to infect Washington's troops. He exposed 300 Bostonians to smallpox then dropped them off at the edge of the city near Point Shirley. Washington provisioned the destitute souls but was under dreadful apprehension that they would communicate the disease to his troops.[4]

Lord Dunsmore engineered a similar situation in Virginia. He offered to free any American slave who agreed to serve in the King's army. One of George Washington's own servants even joined what became known as the Ethiopian regiment. The plan exceeded Lord Dunsmore's expectations. Within a month he had doubled the size of his army to 1,000 soldiers. He packed his regiment into a ship and cruised up and down the Chesapeake raiding plantations, burning storehouses and inducing more slaves to join his ragtag army. The only thing that finally stopped the Ethiopian Regiment was camp fever and probably smallpox that decimated the army aboard Dunsmore's overcrowded vessels.

But George Washington had an ace up his sleeve. Many well-connected colonists had started to experiment with a procedure called variolation. Variolation consisted of simply exposing someone to smallpox in the hope that the person would survive and gain immunity. The well-known minister Cotton Mather had first heard about variolation from his African slave. The servant remembered that, as a young boy in Africa, he had been visited by a numinous old woman who come into his village and scratched all the young children's arms with sticks and then inserted smallpox pustules into the fresh lacerations. The treatment had given the young boy a mild case of smallpox but had effectively immunized him for life.[6]

Cotton Mather had been so impressed with the story that he convinced his good friend Dr. Boylston to start variolating people in Boston. It had not been an instant success. Mobs threw rocks through Mather's windows and rival ministers claimed that Mather and Boylston were "in distrust of God's overruling care."

Massachusetts eventually outlawed the practice but it continued to be popular among people wealthy enough to be able to travel to Philadelphia where the month-long procedure was still legal but expensive.

One of the Bostonians who traveled to Philadelphia to be variolated was John Adams. He wrote about spending several weeks laughing with his brother as they gagged down the Indian potion Ipecac and vomited up their meals. Part of the cure involved ingesting mercury, which loosened the patriots' teeth. Evidently Adams felt the procedure was worth it. He later wrote that probably the main reason he was sent to the Continental Congress was that he was the only person in Massachusetts who had been variolated and would thus be safe among all the people walking around contagious in Philadelphia.[7]

In fact nobody could provide a logical explanation for why the practice worked. Viruses and bacteria had yet to be discovered and there were no concepts to explain immunity. It was easy to see why religious leaders found it blasphemous to deliberately infect someone with a disease, yet a few empirically minded people had seen it work. One of them was George Washington.

Washington had contracted smallpox as a young man when he had traveled to Barbados in order to help his brother Lawrence overcome consumption. Upon their arrival, Washington wrote, "a Mr. Gedney Clarke invited us to dinner. We went-myself with some reluctance as the smallpox was in his family and his wife was much disposed. A fortnight later I was strongly attacked by smallpox." [8]

Though Washington survived the disease, he lived with facial pockmarks for the rest of life. However he also gained a great deal of knowledge and respect for the deadly disease. That knowledge would hold him in good stead when he had to make one of the most momentous decisions in medical history — the decision whether or not to variolate his troops.

George Washington had initially forbade soldiers from variolating themselves for fear it would put other soldiers at risk. Despite his orders many soldiers disobeyed. Ethan Allen even counseled his soldiers to obtain smallpox pustules from sick comrades then lacerate themselves under the fingernails or in the thigh to avoid detection. [9]

By 1777, however, Washington wrote that he "feared smallpox more than the sword of his enemies" and indicated that Lafayette was pushing him hard to variolate his troops. [10]

But it was Dr. Shippen; George Washington's personal physician and Dr. Sutton who finally convinced Washington to variolate his army. Sutton was a successful doctor who had abandoned his lucrative variolation practice in Philadelphia in order to join the wartime effort.

After the decision to variolate had been made, the entire operation had to be carried out in secret, often over the objections of men who didn't want to undergo the grueling, sometimes dangerous procedure. During the height of the variolation campaign in March 1777, there were only 7,000 American soldiers healthy enough to fight. If the British had attacked then, the war would have come to a rapid and unfortunate conclusion.

The logistics were also formidable. Variolation centers had to be set up in Virginia, Pennsylvania, New York, New Jersey and Connecticut. Doctors had to interview every enlistee to determine if they had ever been exposed to smallpox. Officers had to commandeer buildings that could secretly house thousands of soldiers during their enforced illness. Hundreds of coughing, hacking, vomiting men crowded into barracks like the Wampanoags in their long houses. There were rumors that the procedure was tried out first on Hessian prisoners of war.[11]

By the beginning of spring, Washington's army had started to slowly recuperate from the forced epidemic. Nearly half of his troops had recovered by the time the British discovered what had happened. By May, over 15,000 soldiers were well enough to continue the war. Washington could finally feel comfortable sending his troops anywhere in the country knowing they were protected from the threat of smallpox. Washington had staked the very existence of his army on variolation. He had gambled and won.[12]

History has a way of remembering the deeds of humans while forgetting the role of germs. But the numbers tell the story. More people died from disease than from combat in most wars including World War I, World War II and The Civil War. It was the same with the American Revolution. By the end of war 25,000 soldiers had died in combat but at least five times that many Americans had died from smallpox.[13]

Washington's decision to inoculate his troops proved to be as important as any strategy he used on the battlefield. It was carried out in secret, under quarantine and over the objections of his many detractors.

But one mystery remains. How did variolation really work? In Patuxet 100% of the Wampanoags died from contracting smallpox but in Valley Forge less than 5% died from variolation. It was essentially the same disease. In one instance a soldier seeing he was to be to be housed with several prisoners with smallpox, simply peeled the pustule off one of his fellow inmates and inoculated himself with a knife.[14] There is still no fully satisfactory explanation for the lighter symptoms and better outcome of smallpox caused by variolation than from the natural method. Was it because virions that landed on the mucus linings of the throat or lungs performed an end run around the immune system?

Was it because the immune system of the blood was able to act more quickly to recognize and thwart the cutaneous form of the disease? Was it because virus from a scab was more attenuated than virions from a person actively contagious with the disease? We still don't have all the answers.

Textbooks would have us believe that medicine marches purposefully along from understanding to cure. In reality it is usually works the other way. A doctor notices that a drug works against a particular disease and it is only later that a scientist discovers why. Aspirin is one of the prime examples.

Considering the lack of knowledge in 1777, it was truly remarkable that George Washington made such a far-sighted decision. It is not an exaggeration to say it was a decision that changed the outcome of the war, the creation of America, indeed the very history of the world. Future Presidents would have to make similar decisions before sending troops into harm's way or even to just protect their civilian populations.

CHAPTER 3
Four Scientists
1796 - 1928

Variolation was all the rage in England as well as in America. In London, it was being championed by Lady Mary Wortley Montagu, the wife of the British ambassador to the Sultan's court in Turkey. In 1726 she had watched Turkish families having their children variolated. The parents were particularly interested in having their daughters variolated because they feared their daughters would be less marriageable if their faces were covered with pockmarks. She wrote to a friend back in London:

"There is a set of old women who make it their business to perform the operation every autumn, in the month of September, when the great heat is abated. People send to one another to know if any of their family has a mind to have the smallpox: they make parties for this purpose; and when they are met (commonly fifteen or sixteen together) the old woman comes with a nutshell full of the matter of the best sort of smallpox and asks what veins you please to have opened. She immediately rips open that which you offer with a large needle and puts in as much venom as can lie upon the head of a needle, and after binds up the little wound with a hallow shell... You may believe I am very satisfied of the experiment since I intend to try it on my own dear little son."[1]

When Lady Mary returned to England she made variolation her cause, extolling its virtues to her friends in the upper classes. Wealthy families paid extra for pockmarked servants to care for their children while they were being variolated. Lady Mary even convinced the Princess of Wales to Inoculate her children —- but only after the procedure was tried out first on six prisoners condemned to death.

News filtered down from the aristocracy to the lower classes. Although the procedure was being used more widely, it was dangerous. People still had the real disease. Some died from the virus, others from contamination. Even today patients die from vaccines tainted with toxins, viral infection or from their own compromised immune systems. Biology is a complicated business. But in the 1790's people were more than willing to pay for variolation, and doctors were more than willing to oblige.

Edward Jenner

One of those doctors was Edward Jenner, a country doctor in Gloucestershire England. Dr. Jenner was also one of those rare physicians who listened to his patients and didn't scoff at the lore of country folk:

"Among those whom I was called upon to inoculate, many resisted every effort to give them smallpox. These patients I found had undergone a disease they call the cowpox contracted by milking cows with a peculiar eruption on their teats. I was struck by the idea that it might be practicable to propagate the disease by inoculation, after the manner of smallpox, first from the cow, and finally from one human to another." [2]

Now here was an idea that didn't make sense — give a person one disease to fight another? Absurd! But in 1796, Dr. Jenner carefully removed some pus from an open pustule on the arm of Sarah Nelmes, who had picked up pox from her beloved cow, Blossom. Jenner then scraped the pus into the arm (or perhaps the rear end, history is contradictory about this point) of the eight-year-old youngster James Phipps. James developed a mild disease, recovered and was forever immune from smallpox.

The procedure became known as vaccination from "vacca", the Latin word for cow. After initial opposition the world embraced vaccination. Soon, so many people flocked to Jenner's country estate that he had to build a special thatched hut where he would vaccinate his patients free of charge. He called the hut his temple of Vaccinia and it still stands today in the Jenner Museum in Gloucestershire. He also purchased a house for James Phipps to honor his rear end action in the brilliant campaign.[3]

But, was Jenner's discovery of vaccination any greater than the first discovery of variolation believed to have been made by an unknown Taoist alchemist somewhere in the mountains of Szechwan? Arguably not. In fact the Chinese healer probably made the greater and more accurate leap of faith. But truth to be told, neither Jenner nor his unknown predecessor, probably a woman, really knew what they were doing. Nobody knew of bacteria, viruses or parasitic protozoa. Nobody knew what really caused diseases or how to cure them. That problem would fall on the shoulders of another country doctor, almost fifty years later...

Robert Koch

One night in 1873, Dr. Robert Koch peered into the long tubes of a microscope his wife had given him for his 28th birthday. Night after night Dr. Koch had trudged upstairs to peer at all sorts of creatures. But tonight he was looking at the black blood of a sheep killed by anthrax.[4]

Anthrax ran rampant in the fields surrounding Wollstein, the German town where Dr. Koch had his medical practice. It was a strange disease. Cows and sheep would live happily for years, but if you put them into a new pasture their heads would start to droop, their knees wobble and blood would trickle out of their nose. A day later the sheep would be dead and farmers would leave them where they lay, believing their fields cursed. Sometimes half of a flock of sheep would become infected but the other half would stay healthy. Sometimes the farmers themselves would break out in boils and die of advanced pneumonia.[5]

"What are these queer little sticks drifting through the globules of blood? Could they possibly be alive? Did one twitch, give a little quiver perhaps?"[6]

Dr. Koch grabbed a pen and sketched some fine ink drawings of the curious little beings. In the rural village of Wollstein Koch was cut off from the intellectual life of the big cities. But he seemed to remember a lecture given in medical school by a professor of anatomy who espoused the unpopular view that diseases were caused by living organisms. Perhaps he had even read an obscure paper by two Frenchmen who had seen similar threads in anthrax-killed sheep and insisted they were the cause of the disease. Insistent or not, without proof theirs was what scientists call a "Just so" story — interesting but without merit.

But, from that night on, Dr. Koch determined that he would prove that the bacilli really were the cause of anthrax — the disease that so disrupted the lives of his country patients. He roamed the fields of Wollstein searching for the carcasses of sheep and befriending butchers who would save healthy fresh sheep's blood for the town's eccentric doctor.

Night after night Koch checked to see that the bacilli were only in the blood of sheep killed by anthrax and not in the blood of healthy animals. He bought mice and taught himself how to dissect them. He learned how to soak a sliver of wood in anthrax blood and slid it into the tiny incision he made in the base of a mouse's tail. He neglected his practice and built a small lab in his office so he could do experiments during the short intervals between patients.[7]

For months Dr. Koch killed mice, harvested their bacteria and injected them into other unfortunate animals. Finally, after the eighth generation of mice had been killed by the eighth generation of bacteria, Dr. Koch was satisfied. The blood of the last victim swarmed with the same bacilli as the blood of the original sheep killed by anthrax. Dr. Koch washed his hands and went out to attend a child suffering from sniffles. He was ready to announce his discovery.

But wait. How could these bacilli, these fragile bits of protoplasm that Dr. Koch had worked so hard to keep alive, move from one sheep to another? Surely they would perish in the cold, in the heat, without water? These questions niggled in the back of the good doctor's brain.

But then one night Koch discovered that one of his anthrax cultures had become infected with microbes. These cultures were another one of his many ingenious inventions. Koch had discovered that he could grow anthrax in the watery fluid found in the eye of an ox. He would put a drop of the culture onto a cover slip, glue it to a microscope slide, flip the whole thing over, and Voila! There it would be: the drop suspended in the well of the slide, isolated from foreign bacteria. But this particular culture appeared to be contaminated.[8] The long sticks and threads he had grown accustomed to seeing had grown dim and tiny bubbles were welling up within their filaments. Koch cursed his luck and put the sample away. No time to clean the slide. He had another patient to see.

A month later Dr. Koch looked at the sample and the damn bubbles were still there. Could they possibly be related to the bacilli after all? He wondered what would happen if he put the sample into another drop of ox eye fluid. He did so and the beads turned back into bacilli. That was it. The beads had to be spores of the bacillus!

Instead of publishing his findings, Dr. Koch launched yet another set of experiments that lasted several more months. The always-conservative scientist had to be sure. But, by the end of the experiments, Koch had his answer. Twenty-four hours after a sheep dies from anthrax, its blood teems with bacilli that slowly transform themselves into billions of hardy spores — spores that can withstand heat, cold, and desiccation. They remain as the carcass of their host slowly rots around them. Days, months, years, even decades later the spores wait in the grass like deadly biological landmines. When another sheep is let into the pasture it trots to the grass that grows so lushly in the nitrogenous wastes of the long dead cadaver. As it snorts and snuffles in the grass it inhales great snoutfuls of anthrax spores. The cycle has started again. The only way to get rid of anthrax, Dr. Koch explained, was to burn the bodies of dead sheep or bury them so deep their anthrax spores could not mingle with the grass and surface soils.[9]

Dr. Koch was awarded the Nobel Prize for his work and the "Koch postulates" are still used today to identify a microbe and prove that it is the agent of a particular disease. Other people had proclaimed that microbes were the cause of disease, but only Koch had proved it. He had launched the world onto a path that could eradicate disease forever. But he had also discovered how to grow the spores of anthrax — spores that would prove uniquely suitable for biological weapons...

So Koch had proved Bacillus anthracis kills sheep and cattle, but he still had not found a vaccine against the fatal disease. That distinction would go to a scientist as different from Koch, as France was different from Germany, and wine was different from beer...

Louis Pasteur

"Roux, Chamberland come down here immediately!"[10]

It was 1880 and Louis Pasteur, already famous for discovering fermentation and pasteurization, was caught up in another one of his grand passions.

As the two beleaguered assistants entered the chemist's lab he commanded them to look at the chickens that they had injected the day before. All the chickens were dead from cholera, as would be expected. But then he directed his assistant's attention to several more chickens that had recovered after they had been injected with an old culture by mistake. The second group of chickens had received exactly the same dose of deadly cholera. But they had somehow resisted the deadly disease and were now quietly eating.[11]

Roux and Chamberland looked at each other and back at their mentor. What was he getting at?

Pasteur spluttered that, due to their faulty procedures he had just discovered everything — how to prevent diseases, how to save lives! He explained slowly that the team had found out how to make a chicken just a little bit sick so it could recover from a disease. Now all they had to do was keep their virulent microbes in bottles until they became attenuated, then inject them into new chickens that could recover and become immune. If they could immunize chickens to cholera then they could immunize sheep against anthrax. If they could immunize sheep against anthrax they could immunize people against any virulent disease. In one stroke their French lab had been able to outdo Koch in Germany and Jenner in England!

At the Academy of Medicine, the impetuous Pasteur boasted that, "I have demonstrated a thing that Jenner could never do with smallpox - and that is, that the microbe that kills is the same one that guards the animal from death."[12] The words of the arrogant chemist, so infuriated Dr. Jules Guerin that the eminent surgeon challenged Pasteur to a duel. Pasteur wisely declined the offer of the eighty-year-old doctor.

But Pasteur had not only belittled the feat of one of the legends of medicine, he had also dared to trespass on the territory of veterinary medicine, a more useful and respected profession at the time. This was too much for Dr. Rossignol, editor one of France's most prestigious horse care journals. Dr. Rossignol challenged Dr. Pasteur to test his anthrax vaccine at a public demonstration in Pouilly-le-fort.[13]

Dr. Rossignol convinced The Agricultural Society of Melun to buy 48 sheep, two goats and several cattle to test Pasteur's theories and the experiment was to be scheduled for May and June of 1881.

On the first day of inoculations Pasteur bowed to the crowd of dignitaries as Roux and Chamberland injected vaccine into the thighs of half of the sheep, goats and cattle. Then all the animals were given a fatal dose of anthrax on May 31.Two days later the crowds returned and there lined up in a row were the dead bodies of twenty-two unvaccinated sheep and two more were staggering around the paddock with black blood oozing ominously from their noses. And what of the vaccinated sheep? They were frisking about in their paddock wondering what all these horse doctors were doing in their fields. On June 3, the London Times carried the story with the size headlines usually reserved for reporting wars: "The experiment at Pouilly-le-fort is a perfect, an unprecedented success."[14]

Even before this, his latest success, Pasteur had turned his attention from science to business. Now he talked and traveled like a modern American Biotech entrepreneur. He met with financiers and established institutes that bore his name. He accepted awards and was wined and dined by the glitterati of Paris and Europe. He set up a vaccine factory on the Avenue d'Ulm. Fermenters frothed and bubbled all day and giant copper vats produced vaccines to save the world's sheep and cattle industries.

But then something went wrong, terribly wrong. Letters poured in from France and Hungary. Sheep were dying, not from anthrax but from Pasteur's vaccine. Most damaging of all was an exacting little report from Germany. Robert Koch had tested the vaccines and found them teeming with scum-forming bacilli and alien cocci, but most galling of all he criticized the great Pasteur for not reporting his failures:

"Such goings-on are perhaps suitable for the advertising of a business house, but science should reject them vigorously."[15]

Pasteur was incensed. Hadn't he, the great Louis Pasteur, discovered pasteurization? Hadn't he discovered the way to rid France of the microbe that threatened the silkworm industry? Hadn't he single handedly saved the French cheese and wine industries? Hadn't he prophesied that science would rid the world of disease when that bothersome little Koch was still in lederhosen?

Truth to be told, Pasteur had grown a little impatient with the plodding pace of science. He now preferred the fast moving world of business and the narcotic high of saving the world. Perhaps he already knew that Koch would win the Nobel Prize while he the lowly chemist would do without.

So the two proud patriots fought throughout Europe. France against Germany, Germany against France, science be damned. By the late 1880s their arena had moved to Egypt where they and their seconds were competing to find the bacteria that causes Asiatic cholera.

For months the two teams of Robert Koch and G. Theodor Gaffky on one side and Emile Roux and Louis Thuillier on the other raced to cut up cadavers, isolate bacilli, and shoot them into scores of experimental animals. It was international competition as compelling to the 19th Century as was the Space Race to the Cold War world. All appeared to be going well until a messenger delivered frightening news to Robert Koch in 1883:

"Dr. Thuillier, of the French Commission, is dead of cholera."[16]

There it was in black and white, the first scientist to be martyred in the race to eliminate disease. The world mourned. Dr. Koch helped carry the casket and laid a wreath at the funeral. Wiping his eyes he said of the flowers, "They are simple but they are of laurel, such as are given to the brave."[17]

So there it was. In a line that started with an unknown Taoist monk and went through Edward Jenner, Robert Koch, and Louis Pasteur, scientists had discovered the germ theory of disease and started to figure out how you could vaccinate people. But one of the drawbacks of a vaccine is that you have to be vaccinated before you come down with a disease. Wouldn't it be equally important to have a medicine that could conquer a disease after you had been exposed? Wouldn't it be useful to have a cure as well as a preventative? To find out more about that strategy we must go to the cluttered London laboratory of a Scottish scientist who seemed to make a career out of making fortuitous mistakes.

Alexander Fleming

In 1928, Alexander Fleming was busy cleaning up his lab before seeing some patients. Two weeks earlier he had been puttering around with some experiments to prove his pet theory that his own nasal mucus could somehow fight bacteria. But it had been a long busy summer and he had been in a rush to close his laboratory before vacation, so he had accidentally left a petri dish out on a laboratory bench. The petri dish had been smeared with staphylococcus bacteria and while he was away the weather had warmed and the bacteria had flourished.

Most scientists would have simply thrown out the petri dish and started over again. But something caught Fleming's attention when he returned. Alien spores had somehow contaminated the sample. He later discovered that these were fungus spores that had wafted into the open window of his lab from the downstairs laboratory of a colleague who studied mushrooms. At the moment Fleming was probably more annoyed with himself for having left the samples out in the open instead of in the laboratory cooler, but there was something curious about the tiny little colonies of fungi. Each colony was surrounded by a clear halo. Something was killing the thick mat of Staphylococcus bacteria.

Now this was something truly amazing. His nasal mucus had been pretty unimpressive at fighting bacteria, but here was a lowly ground fungus killing bacteria right in front of his eyes. Fleming had discovered penicillin, the first of a long line of antibiotics.[18]

....

So there they are, the fathers of modern medicine: Edward Jenner the open-minded doctor humble enough to listen to country lore, Robert Koch the dedicated scientist willing to work long hours to elucidate the exact cause of disease, Louis Pasteur the swashbuckling entrepreneur eager to build an empire around the lucrative business of saving lives, Alexander Fleming the lucky Scottish doctor who learned from his mistakes - plus the quiet Thuillier who lost his life trying to save others.

All of them good scientists, their work immeasurably important. All were showered with prizes and the adulation of a thankful world. We will see their types emerging again and again in the annals of biomedicine.

It had taken them over a hundred years, but these pioneers had discovered our two primary defenses against infectious diseases: vaccines against viruses and antibiotics against bacteria. Science still didn't know exactly how they worked, but it didn't really matter. One by one vaccines and antibiotics started to control anthrax, smallpox, diphtheria, tuberculosis, cholera, typhoid, and typhus.

It looked like all infectious diseases would soon be a thing of the past. But somewhere along the line, the urge to use this powerful new science of biology to save the world from disease fell victim to the dark side of human nature and its equally powerful urge to use science as a means to gain advantage over one's sworn enemies.

CHAPTER 4
Black Measles
Bitterroot Mountains, Montana
1866

Ranchers in Montana had been plagued by black measles since the early 1800's. But when they started to clear trees to raise more livestock the disease became particularly common on the shady west slope of the Bitterroot Valley.[1]

As early as 1866 a local doctor noticed a tick embedded in the flesh of one of his black measles patients and speculated in his clinical notes that it was the cause of the disease. So in 1906 the U.S. Public Health Service recruited Howard T. Ricketts from the University of Chicago to solve the black measles problem. Through laborious work he discovered that the fever was caused by a tiny Rickettsia organism that was halfway between the size a bacterium and a virus.

Ricketts died while researching Typhus, another rickettsia disease, in Mexico. His death marked the first of several researchers who would give their lives studying what we now call Rocky Mountain Spotted Fever, or *Rickettsia rickettsia* named for Ricketts who first discovered the pernicious new organism.[2]

But today, Howard Ricketts is less known than his distant relative marine biologist Ed Ricketts who John Steinbeck made famous as "Doc'" in Cannery Row. In the book as in life, Ed was notoriously circumspect about his family connections back in Chicago.

But when the governor of Montana's daughter and son-in-law died from Rocky Mountain Spotted Fever, the U.S.[3] Public health Service set up a lab in the abandoned Bitterroot schoolhouse. There, researchers discovered that if they crushed up hundreds of wood ticks into a juicy mush and added carbolic acid they could make a live organism vaccine against Rocky Mountain Spotted Fever. It was dangerous but better than nothing.

Two of the researchers died from working in the schoolhouse that was full of ticks in all stages of development, because nobody knew the risk of working in a contaminated laboratory. Today we would call that a laboratory accident, which had occurred in a biohazard area. The schoolhouse was finally destroyed and replaced by the U.S. Health Service's Rocky Mountain Laboratory in 1924.

In 1948 Willy Burgdorfer fell in love with ticks when his mentor Rudolf Ceiby gave him a petri dish full of sand. Nothing happened until Ceiby shook the dish and twenty soft-bellied tampans boiled to the surface.[4]

The ticks became the center of the young Swiss scientist's world. Professor Ceiby would airmail him samples of ticks from small villages in Africa where Ceiby was trying to find the locus for relapsing fever. Willy would tear open the packages, dissect the ticks under his compound microscope and telegraph back whether he found the borellia spirochete responsible for the debilitating disease.

Ceiby was the scion of the wealthy Swiss family who had started the J.R. Ceiby textile dye manufacturing plant near the Swiss borders of both France and Germany. His family became adept at selling to all corners from their haven in neutral Switzerland.

They produced the polar red dye that Hitler used as the background for his Nazi swastika flags and Rudolf carried on the family tradition by placing promising young Swiss scientists into institutions that supported the U.S. biological warfare program.[5]

So when Burgdorfer finished his dissertation Dr. Ceiby pulled him and another student aside and told them about two choice research positions. One was testing the Ceiby company's new insecticide DDT in Sardinia. The other was working on ticks in the Bitterroot Mountains of Montana.

Both young men lusted after the food and beaches of Sardinia so Dr. Ceiby flipped a franc and it came down with the Swiss warrior goddess Helvetica face down on the floor. Willy had lost and was off to the Rocky Mountain Laboratory in Hamilton, Montana.[6]

Burgdorfer didn't know it then, but he had landed at the right place at the right time. The lab maintained a collection of 13 different tick species that harbored 35 strains of spirochetes.

Willy married, settled down and started working on developing a better vaccine for Rocky Mountain Spotted Fever and eventually other tick borne diseases.

CHAPTER 5
The Spanish Flu - Boston, Massachusetts
1918

As the coronavirus spread across the face of our planet in 2020, researchers looked back at the Spanish flu that was the most deadly pandemic in modern history.

But the first thing they found was that the 1918 Spanish flu was preceded by the 1890 Russian flu. It is quite likely that millions of older people didn't catch the Spanish flu because they had already gained acquired immunity from the Russian flu.[1]

This was of course before our present age when scientists monitor the Far East to see what flu has emerged out of birds or mammals so they can develop a vaccine to combat the flu they expect to infect the West during the oncoming flu season.

Flus are characterized by their proteins. So the 1918 Spanish flu, the 1976 and the 2009 swine flus were all H1N1 influenza viruses, but had different characteristics and virulences.

There were three theories for where the Spanish flu came from. None of them involved Spain. This was because World War I censors forbade journalists from writing stories about the flu on the front lines in France and Germany, so all the early stories about the flu came out of neutral Spain, and it paid the price.

The first theory had the flu coming out of Kansas like a tornado heading for Oz. The second theory had the flu emerging from a British staging area and overcrowded military hospital in Etaples France. The base housed a poultry farm and piggery and over 100,000 soldiers passed through the camp every day on their way to the World War One battlefields.

Other authorities have argued that the flu started in Northern China in 1917 and was spread to Europe when 90,000 Chinese laborers were deployed behind the British and French lines where the infection was picked by soldiers and eventually sailors.

We do know that in the waning days of August 1918, the city of Boston was awash in good cheer. World War One was about to end. Husbands and sons would soon be coming home and the Red Sox were in contention to win the 5th world series with the help their Sultan of Swat Babe Ruth. Plus students would soon be flooding into the city's colleges, as another school year was about to begin.[2]

So nobody paid much attention when a handful of sailors returning from Europe fell terribly ill aboard their Naval vessel docked at Boston's Commonwealth Pier. But within a few weeks thousands of people were suffering from wracking coughs and pneumonia brought on by cytokine storms caused when their bodies released an overload of cytokines and histamines to fight the infection. Apparently, somewhere near Boston, the disease had mutated into a far more virulent form than what had appeared on the Kansas military base.

By September thousands of soldiers and civilians were infected. Schools and public gatherings were being canceled and hospitals were filled way beyond their capacities.

One of the hardest hit places was Fort Devens just outside Boston. It was so overcrowded with sick soldiers that the Surgeon General sent four of the country's leading physicians to investigate.

One of the doctors, Dr. Victor Vaughan, recalled seeing, "Hundreds of stalwart young men in the uniform of their country coming into the wards of the hospital in groups of ten or more. They were placed on cots until every bed was full, yet others crowd in. Their faces soon wear a bluish cast; a distressing cough brings up the blood stained sputum. In the morning the dead bodies are stacked about the morgue like cord wood." [3]

In areas like the town of Ipswich, the state took over the local hospital and set up tent cities in the center of town. Schools, churches, bars, billiard halls, bowling alleys, why even opera houses were all closed shut.

To this day people in Ipswich insist that keeping their windows open at night and the open-air therapy practiced in the tent cities saved their grandparents from coming down with the flu.[4]

Meanwhile the newly mutated and far more virulent disease swept across the country and the world; ultimately killing 500,000 million people and reducing the country's life expectancy from 51 years in 1917 to 39 in 1918, 12 years in total.

But the disease then rapidly petered out and on October 21 Boston officially ended its quarantine. Schools and churches opened. People came out of their homes and the Boston Globe wrote, "A mighty effort will be made everywhere and by everyone today to lift and set aside the sadness which the epidemic has brought to Boston." [5]

In many ways the tragedies that befell Boston and other East Coast cities like Philadelphia gave cities like St. Louis the time and warning they needed to, in today's words, adopt social distancing strategies and flatten their infection curves.

The result was that Boston suffered 6,000 deaths out of the 31,500 people who came down with the flu in two short months but only 1,700 died in St. Louis in the following four long months.

Researchers would use those experiences to justify flattening the curve during the 2020 Coronavirus epidemic.

CHAPTER 6
Two Planes
The summer of 1942

It was a warm sunny day in the summer of 1942. The sun shone brightly on Guinard Island off the Northwest Coast of Scotland. A farmer and his dog walked over the heather-covered outcrop and down along the shore where the dog snuffled idly through the glistening moist wrackline. Suddenly the dog paused, dug excitedly and quickly unearthed the body of a recently dead sheep.[1]

On the mainland, two days later, several more sheep came down with anthrax near the village of Aulthea. It was the first time the British public had any inkling of the secret research being carried out at Porton Down just up the road from Stonehenge.

A month before, a Wellington bomber had dropped a 25-pound bomb near 60 sheep tethered to a post in the center of Guinard Island. The bomb had been loaded with thick brown slurry of concentrated anthrax spores. The test was a huge success. All the sheep had died and the island was contaminated for fifty years. At least once, anthrax-laden carcasses washed out of their shallow graves on Gruinard Island and floated to the mainland, where they infected several flocks of sheep. [2]

Local farmers were curious why government officials were so quick to reimburse them for their losses. Winston Churchill was said to be thrilled with the research at Porton Down, "I don't see why the devil should have all the best weapons." [3]

It was not until 1986 that a British firm was finally paid half a million pounds to decontaminate the island. Teams in biohazard suits pumped 280 tons of formaldehyde diluted with 2000 tons of seawater onto the island and removed much of the topsoil. After the decontamination a flock of sheep was brought to the island to graze on the newly cleansed grass, but a local archaeologist, Brian Moffat remains skeptical: "I would not go walking on Guinard Island. There is no reason to suppose the anthrax has not survived. It is a very resilient and deadly bacterium." [4]

Moffat was only concerned about walking on the island. We have yet to see whether an area really can be cleaned up after biological contamination.

....

Halfway around the world, another plane spiraled out of the sky during the same summer of 1942. This was a Japanese plane that circled listlessly over the hamlet of Congshan in the eastern Chinese province of Manchuria. Peasants wading through rice paddies paused to straighten their backs and look up curiously as plane's motor droned through the heavy summer air. It was probably just one of those Japanese pilots out for a morning ride.

The plane then banked over a quiet Buddhist Temple, surrounded by ornately decorated homes whose brightly colored enamel paint glistened in the early morning light. It lumbered over another group of peasants laughing and gossiping as they waded through more rows of bright green shoots of rice, already weighed down with grain. They looked up and watched as smoke began to stream off the plane's tail. Convinced that it was not dropping bombs, the men laughed and returned to their labors. The smoke dissipated quietly over the land.

A week later rats started coming out of people's basements in broad daylight. They emerged from burrows at the edge of the rice paddies. They swarmed down streets, drank water from fetid sewers, and died by the thousands. Now the fleas had no place to go, but they soon found the legs of people working in the paddies or sleeping on their futons. People started scratching their arms and legs until black patches appeared and they came down with fever. Soon twenty people a day were dying and fear gripped the village.

But hope was on the way. A group of Japanese doctors had miraculously appeared in time to save the village. They converted the Buddhist temple into a treatment post and went about their business. One of the afflicted villagers was Wu Xiaonia who had spiked a temperature of 104. One night, in desperation, Wu climbed the stairs leading to the temple. Villagers could hear her pleading for medicine and water. A young Japanese doctor met her at the door. He seemed kind but kept looking at the other doctors. He led her to a bed and asked her to undress. She did as she was told and he secured her to the bed with a large leather belt. It was only when Wu saw the scalpel that she started to scream.

The doctor was upset but continued with his work. He cut Wu open from her chest to her stomach. Wu saw the brightness of her spleen, the color of her liver; her face was contorted in horror and pain. How could this happen? The villagers heard her unimaginable screams and calls for help. The doctor had a hood placed over her head but the villagers could still hear her dying gasps and moans. But finally it was quiet. Other doctors crowded in. They were eager to see how well the plague had spread through her heart, liver and spleen.

The young doctor asked one of the veteran surgeons why they couldn't use an anesthetic. The senior doctor explained that they wanted to see how well the plague had grown in the young woman's living organs. Anesthesia would have given them unreliable data. He counseled the young man that he would soon get used to it. "It was a war and you had to do whatever it takes to win." [5]

A year before, Japanese researchers had injected a Chinese prisoner with the bacterium *Yersinia pestis* and returned him to his 40 by 50 inch animal cage in Ping Fan Manchuria. He could not stand up or lie down but lived there while the plague replicated in his blood. Six months later the doctors chloroformed him lightly then removed all his blood from an incision in the main artery of his thigh. He was still awake and convulsing during the drawn-out exsanguination.

The blood had then been fed to fleas raised in the special flea nurseries of Ping Fan. A slurry of billions of swarming fleas had then been loaded into the tank that had poured the dreaded disease onto Wu Xiaonai's village in the summer of 1942.

The destruction of Wu's village was not the only success of the Japanese biological warfare unit lead by General Shiro Ishii. He had established the biological weapons unit in Manchuria that Japan had occupied since the settlement of the Russo-Japanese war. At its peak the unit was housed in several complexes in Chanchung and Nanking. But the largest was in Ping Fan. It boasted 76 buildings including dormitories, laboratories, barns for housing test animals and prisons for holding human subjects. It covered six square kilometers and employed 3,000 doctors, scientists and troops. Every month the complex could produce 660 pounds of Plague, 1,400 pounds of anthrax, and 2,000 pounds of typhus, as well as raise hundreds of pounds of fleas in 4,500 special nurseries. Throughout World War II the unit carried out biological tests and attacks on troops from China, the Soviet Union and the United States.[6]

In 1944 General Ishii loaded some of his best specialists into a submarine and sent them to Saipan. They had enough equipment to prepare a flea-filled porcelain bomb that the Japanese planned to drop on allied troops as they invaded the island. Fortunately the sub was sunk before it reached its destination.[7]

But General Ishii's crowning achievement was to be "Cherry Blossom at Night." The operation called for loading a folded wing airplane into a submarine and dropping it off the West Coast of the United States. There, a kamikaze pilot would use the plane to pour plague fleas over the city of San Diego. The operation had been planned for September 1945. However, a month before the operation, Soviet troops swept through Manchuria, capturing the biological weapons unit. San Diego had been saved, but humanity had decisively entered the world of biomedical mass destruction. The age of black biology had begun.

CHAPTER 7
Black Ops
1953-1962

Shortly after the end of the Korean War, the biological laboratories in Fort Detrick Maryland established a program to study using ticks and fleas to deliver disease organisms to enemy soldiers and animals.

The advantage of ticks was that they injected disease organisms directly into soldiers' bodies and the ticks would remain alive keeping an area dangerous for a long time.

About the only person who voiced concern about such a program was General Eisenhower, who queried what would happen if such agents blew back onto friendly forces. His concerns were ignored in the enthusiasm for this "humane" new way of incapacitating an army without causing undue deaths.

The program was managed from Fort Detrick in the horsey countryside of rural Maryland, outside Washington D.C. It also included research on animal diseases at a lab on Plum Island in Long Island Sound off the Connecticut shore, and at a facility for mass-producing biological agents in Pine Bluff, Arkansas.

The army recruited scientists such as Willy Burgdorfer to staff the effort, often funding them through the U.S. Public Health Service to conceal their real employer. In all over 13,500 civilians were employed in the U. S. biological warfare effort.

The affable Burgdorfer became famous for using Swiss watch making tools to dissect ticks and for perfecting techniques to force-feed ticks with organisms that caused diseases like, Q fever, tularemia, Western Equine Encephalitis and rabies. When Willy found a microbe and tick that thrived together, Fort Detrick would add it to its list of potential weapons.

A cocktail of three of these agents were stored in an Olympic swimming pool sized tank in Arkansas ready to be sprayed over Cuba to incapacitate Castro's army in the event the United States decided to invade its Caribbean neighbor. The agents would cause vomiting, nausea and extreme photosensitivity. The last thing a Cuban soldier would want to do was pick up a gun to protect his island.

Even though Burgdorfer's ticks were raised under laboratory conditions, accidents did happen. One time Willy sent what he believed to be disease free ticks to a client, but when one of his technicians fell ill with relapsing fever he had to send the recipient an urgent message to destroy the ticks as well as the container they came in.

One of Willy's top priority projects was to mass-produce rat fleas with the *Yersinia pestis* plague organism. The fleas were tested at the Dugway Proving grounds in Dugway Utah.

The project worked perfectly. A military plane dropped cluster bombs filled with 670,000 fleas which fell on the ground where 177 of them made their way to 47 guinea pig cages arrayed around the drop site.

The experiment was reported to be a big success except that one of the bomblets malfunctioned and fleas bit the pilot, bombardier and observer. The project had lived up to its code name, "The Big Itch."[1]

CHAPTER 8
The Test; Dugway Proving Ground
Tooele County, Utah
July 12, 1955

It was almost midnight in the Utah desert. Stars shimmered with vivid intensity. A slight breeze chilled a young man cloaked in a regulation army blanket. He pulled the blanket closer and looked at the cages of monkeys and guinea pigs arrayed around him.[1]

One of the monkeys looked at him impassively. The young man shifted nervously. He had been waiting all week, but now he was not sure he wanted to go through with this test.

At least it was better than at his home base in Fort Detrick, Maryland. There the human subjects had to breathe through a hundred-foot rubber hose that snaked out of the "Eight Ball" a four-story high explosion chamber. The volunteers could feel the shock when the bombs exploded, but they could not feel the miasma of invisible germs that swept deep inside their lungs. At least out here in the desert you could enjoy the cool night air before being bombarded with infection. He wondered if any of his buddies were asleep. He knew they too were out there alone, in the dark, spaced a tenth of a mile apart.

Finally the siren blew. The young man climbed a platform and faced into the wind trying to remain calm. The monkeys started to scream and shake their cages. The guinea pigs sniffed the evening air. Half a mile away sprayers were spewing a fine mist of germs into the quiet desert breeze. [2]

Two weeks later the young man was back at Camp Detrick with chills and temperature close to 103 degrees. His head throbbed and his vision had grown blurry. This was Q-fever, one of the new incapacitating agents of which the army was so proud. Toxins, rickettsiae, and viruses had replaced the "mouldie oldies" of earlier years. Mouldie oldies were plague, anthrax and tularemia; the bacterial diseases that had gone out of fashion with the advent of antibiotics during World War II.

The young man was a conscientious objector, one of almost a thousand Seventh Day Adventists who had volunteered to be dosed with biological weapons in lieu of combat duty. Doctors monitored the volunteers and gave them antibiotics before the diseases became lethal. Animal test subjects were less fortunate.

....

Fort Detrick experts released biological agents over St Louis, Minneapolis, and Winnipeg. They shattered bacteria filled light bulbs in the New York subway system. The germs were supposed to be noninfectious bacteria that only affected plants, but after naval vessels sprayed *Serratia marcescens* into San Francisco Harbor eleven people ended up in the Stanford University Hospital. All the patients were diagnosed with serratia infection, including one who died. In all 800,000 people had been exposed.[3] Could government officials justify losing a few innocent lives in order to prepare the nation for biological warfare? Could they justify denying the loss to protect their secret work? Evidently so: this was The Cold War.

The United States has always harbored a deep-seated ambivalence about biological warfare. In 1942, Franklin Roosevelt had lambasted Japanese and German biological weapons as "terrible and inhumane", then quietly started America's own secret weapons program under George Merck, president of Merck pharmaceuticals. It was housed in Camp Detrick, an army facility in Maryland's Catoctin Mountains fifty miles northwest of Washington. The camp was upgraded to Fort Detrick in 1956. [4]

From the beginning, American military leaders were dubious about the worth of biological weapons. They felt they were dirty and unpredictable. They killed innocent people and could blow back and infect your own troops. Unlike the Soviet Union, the United States had never been invaded by a foreign country and prior to September 11, 2002 its continental shores had never experienced a large-scale foreign attack. Besides, the United States was the only nation that had ever used nuclear bombs — bombs that had ended World War II in a dazzling double flash. The military knew it already had the ultimate weapon.

Even the practitioners of biological warfare had their doubts. In 1949 Theodor Rosebury, one of the pioneers of biological warfare, wrote *Peace or Pestilence,* a book that argued that biological weapons had no military value. He suggested that the United States should turn its expertise away from making biological weapons and toward the new war to eradicate infectious disease. His advice was ignored. [5]

In the mid-Fifties, shortly after the Dugway tests, Fort Detrick turned its attention to viruses. Unlike bacteria, viruses were unaffected by antibiotics. A virus was so small it could evade the body's immune system by slipping into one of its victim's own cells. There it would hijack the cell's nucleus and force it to make copies of itself. As millions of new viruses burrowed out of infected cells, victims would suffer chills, fever and nausea; the symptomatology of viral infection. Such treachery made viruses doubly dangerous.

Of course you could use vaccines against viruses, but people had to be vaccinated before exposure. To military planners this feature made viruses more effective because you could inoculate your own troops to make them invulnerable to secondary infection.

But viruses had another advantage. Like the flu or a common cold; many viruses only incapacitated their victims, making them miserable but still alive at the end of a week. So you could sell viruses as humane weapons that not only saved lives but tied up twice as many facilities as combat deaths. The Kennedy Administration found these properties particularly attractive as they sought ways of defeating their longtime nemesis Fidel Castro.

By 1956 the United States had already started construction on building XI002, a ten-story virus production facility in Pine Bluff, Arkansas. The location had the hidden advantage that it was in the center of America's chicken belt. This meant eggs could be delivered to the facility without attracting undue attention. Every day thousands of eggs would be loaded onto conveyor belts to trundle past technicians who inoculated them with Venezuelan equine encephalitis. The VEE was mixed with *Coxiella burnetti*, the richettsia germ that causes Q-fever. Finally the technicians would blend in a healthy measure of Staphylococcal enterotoxin B, SEB, the bacterial toxin that causes acute food poisoning. Voila, they had concocted a cocktail of invidious almost lethal germs. They made hundreds of gallons of the witch's brew, more than enough to fill an Olympic sized swimming pool.[6]

The cocktails were for the Marshall Plan, an ironically named secret mission to invade Cuba. The plan called for having tanker jets fill up with germs at Pine Bluff, then fly a series of sorties over Cuba. There they would spray a slurry of the germs into the trade winds that would waft gently over the Caribbean island. Within three hours, victims on the ground would be hit with a triple whammy. First they would come down with fever, chills, and muscle pain, hallmarks of severe Staphylococcus food poisoning. Five days later, just before the SEB wore off, they would be hit with the diarrhea, nausea and extreme temperatures of Venezuelan equine encephalitis. Ten days later they would contract Q fever with more fever, facial pain and hallucinations. Finally after three weeks of suffering the victims would recover to find their country occupied but their infrastructure still intact. The only drawback was that 70,000 people, mostly infants and the elderly, would have died. But all the Cuban soldiers would be alive, because they were young and healthy, and the invading American soldiers because the germs were not contagious from person to person. What a weapon. It saved soldiers' lives, spared military targets, and only killed innocent victims!

Say you were a military commander facing an enemy holed up in remote caves in some place like Laos or Afghanistan. Wouldn't you want to be able to achieve your objectives without firing a shot? Soviet commanders were said to have tried biological weapons in Afghanistan in the 1980s — presumably with little success. [7]

It was President Eisenhower who put his finger on the illogic of such thinking. It was at a military briefing at Fort Detrick during the waning days of his administration. The Pentagon's chief scientist, Herbert York, had just opened the meeting promising that biological weapons would "open up a new dimension of warfare." General Lyman Lemnitzer had extolled the revolutionary new weapons as humane alternatives that could be "as mildly disabling as influenza or as deadly as atomic bombs."

The meeting grew silent and all eyes turned to President Eisenhower who had more combat experience than anyone in the room. The former general thanked the military for its informative briefing and congratulated them on their incapacitating agents. But he had one question. What if the enemy didn't think that the use of an incapacitating agent was such a humane act and retaliated with lethal force? [8]

CHAPTER 9
Over Cardenas, Cuba
1962

In 1962 a twin-engine Fairchild C-135 flew over the quiet waters of the Caribbean on its way to the sugar cane fields of Cuba. Aboard it were two pilots, a bombardier, an observer and a new young recruit.

The pilots wore the red white and blue uniforms of Air America, the sham airline the CIA uses for covert operations.

As they approached the island the co-pilot shouted over his shoulder,

"OK, we're getting close."

The recruit ripped open the two unmarked boxes in front of him.

"What the fuck!"

The boxes were crawling with thousands of ticks. The recruit threw them out the open door and slammed it shut.

A few days after another ground-based operation on Cuba, the recruit's son came down with a fever that spiked at 105 degrees. He was rushed to the nearest hospital where doctors wandered in and out of his room at a total loss to explain what was ailing the young child.

Finally a young resident happened to be making rounds and piped up.

"I used to work in a tropical medicine clinic in Cuba, and I have seen this disease before. I know how to treat it."

After his son was cured the recruit asked his commanding officer whether the young boy's illness could have been related to the recruit's mission over Cuba.

"I can't give you any details but you really need to burn all the clothes you were wearing that night. Burn everything!"[1]

The young man learned later that Operation Mongoose, designed to incapacitate Cuban sugar workers during the cane harvest, had been cancelled because Cuba's erratic winds made it too difficult to drop such a payload on the cane fields with any kind of accuracy.

Were they Willy Burgdorfer's ticks? If so, he had come a long way from being an idealistic young scientist dedicated to saving people's lives from infectious diseases.

By 1963 President Kennedy was also getting uncomfortable with biological weapons and invited Matthew Messelson to Washington to offer advice on biological weapons. Dr. Meselson was a well-connected Harvard biologist, who enjoyed dabbling in international affairs. Meselson argued that the United States should not pursue biological warfare because ultimately it would only serve to put cheap weapons into the hands of dictators of undeveloped countries. The best strategy for a wealthy nation like the United States to pursue was to keep warfare as expensive as possible. The argument was politely ignored, but Meselson did not give up.

During the Vietnam era the U.S. biological warfare program went totally off the rails. No hair-brained idea was too crazy to try. Over 280,000 radioactively marked ticks were released in Virginia and Montana alone.

Bioweaponeers had not inoculated the ticks with diseases but as Willy Burgdorfer said, "There is no such thing as a clean tick. They are all little eight-legged cess pools."

Simulant bacteria were sprayed over the San Francisco area and released in the New York subway system. The simulants were supposed to be innocuous but susceptible people got sick and some died a few days after the experiments.

Finally, in 1968, more than 2,000 sheep were killed in a chemical weapon experiment that went wrong and could have wiped out Salt Lake City if the wind had been blowing the other direction.

This was even too much for Richard Nixon who needed to maintain support for the Vietnam War.

That same year, Henry Kissinger had bumped into his old friend Matthew Meselson at Logan airport in Boston.

The two academics had worked together at Harvard. Kissinger asked Meselson what he thought the United States should do about biological weapons.

"Let me think about it. I will write you some papers."[9]

This time Meselson's arguments hit pay dirt. President Nixon needed a dramatic gesture to counter public opposition to the war in Vietnam. So, on November 25, 1969, at a press conference at Fort Detrick President Nixon made the startling announcement: the United States would renounce the use of biological warfare and work on a treaty to ban them from the face of the earth. He called biological weapons "immoral and repugnant," almost the same words used by President Roosevelt thirty years before.[10]

So the United States had decided to end it all - the tests, the accidents, the preparations for biological warfare — but it would not succeed in ending the ambivalence.

There were still people in the audience who felt that the best way to deter the threat of biological weapons was to be able to respond in kind. They believed it was foolhardy to throw away such potent weapons when the chances were good that they would one day be used against the United States.

Some of these people knew how to make biological weapons and had access to classified information. Did they feel disdain for people who they felt were blissfully unaware of the threat of biological warfare? Did one of them become more isolated but more sure of his convictions? Would he become so sure of his convictions that he felt he could justify any action to wake up America — even if it involved the death of innocent victims as had happened in San Francisco in 1950, or with anthrax in 2001?

Only time would tell. For the moment it didn't matter. In 1972, more than a hundred nations signed the Biological and Toxin Weapons Convention. The world had rid itself of an entire class of weapons ... or so it thought. [11]

CHAPTER 10
The Research Paper
1973

Biological warfare was the furthest thing from Kanatjan Alibekov's mind when he entered medical school at the Tomsk Medical Institute in 1973. He had been planning to go into military psychiatry until he took Colonel Aksyonenko's course in epidemiology. The kindly professor had taken a liking to the bright young student from Kazakhstan and had asked him to write a paper on an outbreak of tularemia that occurred just before the 1942 Battle of Stalingrad. [1]

Alibekov became absorbed with the project. Night after night he pored through the 25-volume History of Soviet Medicine in the Great Patriotic War. He tracked down obscure medical journals written at the time of the outbreak. But the information he found in these different sources was contradictory.

The first victims of tularemia were on the German side of the front line. So many of the German Panzer drivers became ill that the ground assault came to a temporary halt. But a week later thousands of civilians and Soviet soldiers came down with the same disease. The History of Soviet Military Medicine in the Great Patriotic War said that the epidemic was natural, but the journals said there had never been such a widespread outbreak of tularemia in Russia before.

Besides, it just seemed strange that so many soldiers had fallen ill on one side of Soviet-German line when the armies were so close together. Why didn't both sides come down with the illness at the same time? One of the journals reported that 70% of the German troops came down with pneumonic tularemia, a rare form of the disease. Suddenly it hit Alibekov: the outbreak had to have been caused by a deliberate dissemination of tularemia germs. Kanatjan rushed to complete a first draft of his paper and knocked expectantly on Professor Aksyonenko's office door.

"Come in, Kanatjan, come in!" Professor Aksyonenko, beaming over the top of the latest edition the Soviet Army's official newspaper asked Alibekov what he had discovered.

Alibekov explained that he had studied the medical records of 1942 and the pattern of tularemia during the Battle of Stalingrad did not seem to suggest a natural outbreak.

The smile that had so warmed Colonel Aksyonenko's face only moments before, congealed into a frozen mask. He folded his paper with studied deliberation and asked Alibekov just what he thought the records suggested. "It suggests sir, that this epidemic was caused intentionally."

"Stop, I forbid you to go on," The silence was palpable. Alibekov stared at his professor in confusion.

"All I asked you to do was describe how we handled the outbreak, how we contained it. You have gone beyond your assignment. I don't want to see this paper until you have given it more thought. And please I want you to do me a favor and forget you ever told me what you just said. Never, ever mention it to anyone else. Do you understand? Believe me, you'll be doing both of us a favor." [2]

Alibekov rewrote his paper without mentioning his discovery. But it didn't matter; the paper had changed his life forever. Military psychiatry no longer seemed so compelling. He changed his concentration to epidemiology and started reading everything he could find about biology and warfare.

Like all students in the Soviet Union, Alibekov had been taught that the Battle of Stalingrad had been the supreme test of the Soviet Union's will to survive. Before the Soviet Union even came into existence Russia had been invaded by superior armies from Europe and Asia. Under Stalin the Soviet Union had grown paranoid that the surrounding capitalistic world was intent on Communism's destruction.

If Stalingrad had been overrun, the Nazis would have eviscerated the Soviet Union's industrial heart and crushed Communism's fledgling experiment. More than a million Soviet soldiers died defending Stalingrad. But in the end, they prevailed. The motherland was saved, and the Nazis slunk back into Europe as had Napoleon's troops centuries before.

The lesson of the Battle of Stalingrad was irrefutable. If your country, and all it stood for, was faced with certain annihilation, you were morally obligated to use any weapon at your command. As a scientist Alibekov had become fascinated with the notion that disease could be used as an instrument of war, that biology could be used to develop the ultimate weapon.

In later years Alibekov learned more about the early use of biological weapons. The Soviet Union had been forged in the bloody crucible of civil war. After the revolution Red and White armies surged across the ravaged face of mother Russia, battling from Siberia to the Crimean peninsula. By 1921 over ten million people had died from bullets, bombs, and famine.

This was the time when Dr. Zhivago and Lara traveled the length of Russia in Pasternak's romantic tale. This was the time when they tumbled into illicit love tending to the sick and wounded. This was the time when they encountered the bawdy witch Kubaricka who sang a magic chant to cure anthrax on a cow's udder, "Terror, terror show your mettle, take the scab, throw it in the nettle." And this was the time when Kubaricka warned young women not to be fooled by the red banner of "The death woman, who nods and winks and lures young men to be killed, then sends famine and plague." [3]

The lessons of the war were not lost on the commanders of the Red Army either. More people had died from typhus than from the actual fighting. The Communist government trained thousands of medical workers and initiated a broad-based system of public health. But in 1928 they also passed a secret decree ordering their scientists to turn typhus into a battlefield weapon.

Thousands of chicken eggs were diverted to the Leningrad Military Academy to be inoculated with typhus germs that were then made into primitive aerosol weapons. Many of the weapons were tested on the prisoners of Solzhenitsyn's "Gulag Archipelago". Typhus was an invincible weapon for almost twenty years until scientists developed a vaccine against the disease during World War II.

The Soviet Union's biological weapons program expanded dramatically after its troops captured blueprints of the Japanese biological warfare complex in Ping Fan Manchuria. Under Laurentii Beriya's watchful eye the KGB used the documents to construct the Sverdlosk biological research complex in the Ural Mountains a thousand miles southeast of Moscow. This was the same city where the Tsar's family had been butchered to end Romanov rule. But Sverdlosk would soon become as famous for its anthrax as for its Romanovs and as infamous as Chernobyl in the murky world of biological warfare.

A year after the Sverdlosk complex was completed in 1946, Stalin was at it again. This time he ordered the construction of the Soviet Union's first smallpox factory in Zagorsk, an ancient cathedral town 40 miles northeast of Moscow. Zagorsk would come into its own in 1959 when a traveler from India inadvertently helped the Soviet Union's smallpox effort by infecting 46 citizens in Moscow. The visitor arrived in Russia with smallpox, but he was not diagnosed for several days. Because he had been vaccinated as a child, the vaccination's effects had worn off. His immune system was still strong enough to protect him from the symptoms of smallpox but not strong enough to prevent him from passing it on to others—potentially thousands of others.

But in the end only 46 people died. Soviet authorities had narrowly averted a major epidemic largely through luck and strict quarantines.[4]

After the crisis, the Soviet Union sent doctors to India to vaccinate its citizens against smallpox. The program was well received and eventually led to the extermination of smallpox on the sub continent. However, the Soviet Union also dispatched KGB agents to India to retrieve samples of the highly virulent strain of smallpox that had almost devastated Moscow. This strain, known as India-1967, became the Soviet Union's principal source of battlefield smallpox. By the 1970's Zagorsk was stockpiling 20 tons of India-1967 a year, enough to fill the fleets of ICBM missiles targeted toward the western world. [5]

Ironically, during the same time the Soviet Union was pulling ahead of the west in black biology, its non-military white biology was being crippled by the politically correct but scientifically disastrous academician Trofim Lysenko.

During the 1920's Lysenko had performed a series of experiments growing wheat in colder and colder conditions. He concluded that organisms could adapt to the environment and pass on these adaptations to their offspring — that nurture was more important than nature. This was politically correct but scientifically absurd. It fit in perfectly with communist doctrine but was in direct contradiction to evolution. In fact it was classic Lemarchianism that had been proved wrong by Charles Darwin in 1865.

But Lysenko had concluded that Darwin was a capitalist apologist and that evolution and genetics were bourgeois disciplines that insulted the proletariat. Simply by growing wheat in colder and colder temperatures you could make wheat that liked to grow in cold climates, simply by raising people in collective settlements you could make people who would put the good of society before the good of the individual. It sounded too good to be true. Indeed it was. If the experiments had been done by an unknown scientist with no political affiliations the scientific method would have prevailed and the results proved wrong. [6]

Unfortunately for science in the Soviet Union, Lysenko was one of Stalin's rapidly rising Communist cronies. He became a leading academician and imposed an iron brand of political correctness on Soviet biology. No one was allowed to publish papers or do experiments in genetics. Not only did Lysenko almost single-handedly ruin Soviet agriculture by ordering crops to be planted where they could not thrive, but he crippled academic biology as well. Soviet scientists were forbidden to subscribe to Western journals or travel to international conferences.

It was during this time that western biology was going through its own revolutionary change. In 1953 James Watson and Francis Crick discovered the structure of DNA. Then, in the 1970s, an ambitious young biologist named Joshua Lederberg discovered that bacteria regularly pass genes back and forth to each other by forming tiny rings of DNA called plasmids that can flow from one organism to another. If this happened in nature, why couldn't it be done in the lab? In 1973 Stanley Cohen and Herbert Boyer proved it could. They spliced a gene for resistance to penicillin from one bacteria into E. coli, the normally benign bacteria that live in the human gut. In effect, they had created a biological weapon. Thus, the very first experiment that ushered in the brave new world of genetic engineering had also shown its potential as an instrument of war.

The lesson was not lost on the few Soviet scientists who could smuggle Western journals into the country. The experiment showed that Soviet biology under the iron fist of Lysenko was woefully behind the West. But to a few scientists it also showed something else. Here was a way to throw off the chains of Lysenkoism. The way to do it was to convince Soviet officials that you could use gene splicing to create new, more invincible weapons, superbugs that could evade vaccines and antibiotics.

Only one scientist had the political power to pull this off. He was Academik Yuri Ovchinnikov, who started extolling the benefits of gene splicing to Leonid Brezhniev and like-minded generals in his private chambers. Gradually the lessons sunk in. In 1973 Ovchinnikov convinced Brezhniev to establish Biopreparat, a civilian agency that would provide the military with advanced research on biological weapons. Biopreparat would prove to be as ambitious and successful as the Soviet program to build the atomic bomb.

Ovchinnikov had successfully rescued Soviet biology from the yoke of Lysenkoism by harnessing it again to Soviet militarism. [7] Like most Soviet citizens Alibekov knew nothing of these secret dealings. He was too busy graduating near the top of his class in medical school — little did he know how much these political intrigues would change his life, little did he know how much his brilliance was already being noticed by those in charge of Biopreparat.

CHAPTER 11
Tightening the Noose;
The Smallpox Eradication Program
1958- 1978

During the same time the great powers were starting their first biological warfare programs, white biologists were on a happier quest: the eradication of smallpox. It was first proposed in 1958 by the Soviet Union, which had a long history of using science to understand and fight viruses. First they built on the work of their early Russian forbears:

In 1831, Elia Metchnikoff was sitting in a humble apartment overlooking the Straits of Messina. It was Christmas and he had just been fired from the University of Odessa, an already frequent occurrence in the young man's mercurial career. But Metchnikoff knew that if he could just get some time to do his own research he could accomplish great things. Accordingly, he had converted the family's small parlor into a functioning marine laboratory. Olga, his wife, had obliged by taking the children to see some performing monkeys so he had the whole morning to observe starfish cells under his microscope. But as he watched the cells move he was struck by an intriguing notion. What if the cells moved to attack and consume microbes? That could protect the starfish from infection.

Metchnikoff rushed outside to collect his thoughts. A rose bush grew beside a tangerine tree the young family had decorated for Christmas. It was all they could afford. But the rose gave Metchnikoff an idea. He stripped a thorn from the bush, returned to the lab, and jabbed the thorn into the starfish. If his theory were correct the starfish cells would swarm around the thorn. He remembered that when people get splinters in their hands; the splinters were soon surrounded with white pus, the remains of dead white cells. [1]

The next morning he rushed back to his microscope. The starfish cells had migrated to the thorn and were now swarming over it, consuming microbes. Metchnikoff had discovered that these wandering cells, not Koch's mysterious humours, were responsible for fighting infection. It was the beginning of our modern understanding of immunology. Eventually we would learn how vaccines can trick the immune system into producing antibodies, but for the meantime, the vaccines were working just fine.

Some sixty years later, another Russian scientist would perform a similar experiment to further unravel the puzzle of illness. In 1892 plantation owners in the Crimea were concerned about a disease that was causing their tobacco leaves to break out in strange mosaic patterns. Dimitri Ivanovski thought the disease was probably caused by a plant bacterium, so he tried to isolate the agent by pouring samples of it through filters. The filters were normally used to isolate bacteria, but the samples of this agent could pass through the filters and still infect other plants.

Ivanovski gradually realized he had discovered a new class of germs; they were viruses so small that no one would actually see one until the discovery of electron microscopes almost a hundred years later.[2] Once again science had proceeded by fits and starts — not in a logical progression.

The Soviet Union stood on the shoulders of these early Russian scientists to help it develop communist-inspired medicine. It sought out and trained thousands of students who formerly would have been excluded from medicine and concentrated on large, government run public health initiatives. One of its early successes was a program to eradicate smallpox in the republics of the Soviet Union. By 1936 the program had succeeded and Soviet officials realized it would be both good medicine and good politics to start exporting Communist-style medicine to the Third World. Soon medical training and vaccination programs became part of the Soviet Union's diplomatic and humanitarian outreach programs. They were well received in Africa, Asia, and Latin America.

Backed by this solid record, the Soviet Union proposed in 1958 that the World Health Organization spearhead a campaign to rid the world of smallpox. Now this was a truly audacious proposal. It was the first time that mankind had stated so clearly its intention to wipe out another species. But this species was so small it had never even been seen and so numerous that a single person could harbor more smallpox germs than all the people on earth.

Experts realized how high the stakes really were: if they missed even a single germ it could roar back with increased ferocity. If the program failed, the world would never again undertake such an ambitious program and vaccination itself might come under suspicion.

But there were reasons for optimism as well. Smallpox does not exist in "nature", it only exists in humans. There were no ticks, fleas, or mosquito vectors that had to be controlled, no birds, pigs, bats or monkeys that could act as reservoirs from which the disease could reemerge more evolved and lethal than before.

Then there was the vaccine itself. It appeared to be uniquely safe because it was based on a disease other than smallpox. You couldn't contract smallpox from smallpox vaccine because ever since Jenner's discovery it had been made from cowpox viruses. Other vaccines used dead or weakened germs so there was always the possibility that a person could come down with the same disease they had been vaccinated against. The worst thing that could happen with the smallpox vaccine is that someone could come down with a case of cowpox. By 1958 the vaccine was 99% effective with no side effects, other than the small cowpox scar that appeared on your left arm. [3]

The campaign was officially started in 1967. It was one of the first efforts to bring together top international scientists to achieve a common goal, and it became a model for later international efforts in space. Dr. Donald Henderson was selected to lead the campaign. He was a charismatic American who attracted cadres of idealistic young specialists in tropical medicine. Teams of them fanned out across the globe to identify and vaccinate populations threatened by smallpox. They would proceed village by village, gradually tightening the noose on the disease, until the last case was corralled and the last germ killed.

Sometimes this was done with little sensitivity. Medicine men were forced to give up their valuable stashes of smallpox scabs, while mostly white Soviet, European and American specialists lorded it over local medical authorities. But the world health teams regarded this as a holy war and they gained a reputation for being able to overcome any famine, war, or bureaucratic hurdle in their headlong quest to conquer their deadly foe.

There were many times when the program almost ground to a premature halt. One of the most difficult periods was in 1974 in Northern India. The area was in the grip of several epidemics of smallpox at the same time it was experiencing runaway inflation. Gasoline was so expensive that airlines and railroads had gone on strike, then the health staff went on strike as well. Dr. Henderson feared that smallpox would jump out of the quarantined areas and spin wildly out of control. At the very last minute he was able to help broker a settlement with authorities and scrounge up enough gasoline to continue the campaign. By May 1975 his teams had isolated the last case of smallpox in India. [4]

Daniel Tarantola was having his own problems. He was head of the World Health Organization team in Bangladesh. Village elders kept reporting to him that a gang of notorious thugs with pockmarked faces was threatening their villages with smallpox. No one dared confront the known murderers for fear of getting killed. Dr. Tarantella had studied in Paris during the 1968 riots but he was not prepared for this. The army and the local police were no help. Dr. Tarantella had to approach the leader of the gang in his mountain hideout, armed only with a syringe and common sense. Finally the leader put down his gun and presented his arm. It turned out to be too late for him: the notorious murderer died two days later, but the local epidemic had been stopped. [5]

In 1975 Donald Henderson flew to New Delhi to celebrate India's first full year without smallpox. He was scheduled to fly on to Bangladesh where victory also seemed to be within Dr. Tarantella's reach. But just before boarding the plane to Dhaka he heard the news. Sheik Mqjibur Rahman, the Prime minister of Bangladesh and the father of its independence, had just been assassinated. The airports were closed and Bangladesh was preparing for civil war. [6]

Dr. Tarantelo was facing more immediate problems. The monsoons had hit just as he was about to close in on one of the last cases of smallpox. Now roads and bridges were washed out, and cities and towns were inundated with water. His vaccination teams couldn't even find the existing cases, let alone the epidemic that was sure to follow. Dr. Tarantelo had to round up his dispirited staff and give them a pep talk worthy of an NFL coach.

"Look, this just means we have to get down to micromanagement. We must look at the trees for now, not the forest. Take it day by day." [7]

The pep talk worked, his staff slogged back into the field and meticulously tracked down each case. Excitement was mounting. No new cases had cropped up in India or Africa; only Bangladesh was left. All eyes were on Dr. Tarantelo's team. The last case was in Chittagong, a city ruled by an arbitrary general known for his nasty disposition. Dr. Tarantelo was turned down flat. His teams could not go into the city to vaccinate it's populace. The epidemic flared and threatened to spread into the surrounding countryside. Under pressure the general finally relented and Dr. Tarantelo's team was able to find and eliminate Bangladesh's last case of smallpox. The team was exhausted but they managed to stay up all night with some of the best champagne known to France. But their celebrations were premature. [8]

The next morning a cable arrived. Smallpox had surfaced on Bhola, a tiny island in the Indian Ocean off Bangladesh. For the third time, the team had to slog back into combat after thinking that it had won the war. This time they waited for several months before breaking out the champagne. But finally Donald Henderson was able to announce that Bhola's three-year-old Rahima Banu had been the last person on the earth to have *Variola major*, the severest form of smallpox. The last case of *Variola minor* was isolated in Somalia in 1976, and the world was officially declared smallpox free in 1980.

· ·

During the height of the Cold War the Soviet Union and the United States had led an international effort to eradicate the worst disease known to mankind. It had taken eleven years to accomplish and cost $300 million dollars, a pittance for such a formidable achievement. [9]

But the most crucial lesson that Dr. Henderson had learned was the importance of sticking to the vaccination program until every last germ had been wiped off the face of the planet. It was a little like taking antibiotics. If you didn't kill that last germ it could return more virulent and resistant than ever. Even when things had looked futile Henderson had managed to squeeze out a few more dollars of funding, and a few more years to finish the job properly.

Unfortunately political leaders had not learned the lesson so well. In 1956 the United States Congress funded a program to rid the world of malaria by 1963. When the deadline ran out the program stopped. The result? Malaria returned many times more powerful than before.

Mosquitoes had evolved resistance to DDT, and plasmodium, the protozoa that causes malaria, had evolved resistance to chloroquine, then quinine and primaquine.

By 1975 the incidence of malaria was 2.5 times what it had been in 1961. China and India had jumped from 1 million cases to 6 million cases each. Humankind had created a more virulent disease than before. Even gin and tonic, the old British remedy for malaria was found to give people Blackwater fever, a man-made disease caused by ingesting too much quinine![10] Such diseases are called iatrogenic, caused by the patient's activities. So in a sense, today's virulent form of malaria is also an iatrogenic disease, as are the world's supply of germs made for biological warfare.

But for the moment the world could rejoice. Smallpox was a thing of the past, and countries could stop vaccinating their children. The United States did so in 1972. But what if smallpox were not such a thing of the past? What about the billions of children who would no longer be vaccinated? Could they face the same fate as Rahima Banu?

CHAPTER 12
"To Inoculate Every Man, Woman and Child."
The Swine Flu Fandango
1976

In 1976, Judy Roberts went to get vaccinated. She had never had a flu shot before, but she had been convinced to get immunized by a barrage of government sponsored TV ads. If Walter Cronkite, Muhammad Ali, Rudolf Nureyev, and President Ford could roll up their sleeves on national television, why shouldn't she? It seemed like the intelligent and patriotic thing to do.

Two weeks later Judy noticed a numbness in her toes that seemed to be creeping up her spine. She joked to her husband, "By Friday, I'll be numb to my knees."

By Friday she was a wheelchair-bound quadriplegic who had just undergone a tracheotomy in order to breath. Though she would eventually be able to get rid of her wheelchair, she would have leg braces and weak hands for the rest of her life.[1] How did it happen?

1976 was supposed to be the year America would finally bury the legacy of Vietnam and Watergate in a yearlong celebration of the nation's birth. It would culminate on the Fourth of July with the Tall Ships sailing majestically beneath New York's newly built World Trade Center Towers. Instead, the year became known in the medical community as the case study for how not to run a vaccination program.

The problems started shortly after Christmas 1975 when five hundred new recruits reported to Fort Dix, New Jersey, for active duty. As the young men mingled in the reception area getting physical exams and shots, they also swapped viruses. Private First Class David Lewis got the short end of the stick. He and 11 other recruits contracted A/New Jersey, which would soon become known as "the Swine flu killer virus." Three hundred other recruits came down with ANictoria, a milder flu strain that had first emerged in Victoria, Australia, in 1975. Most of the sick soldiers rested in the base dispensary, but Private Lewis ignored his symptoms to join an all night hike through the frozen Pine Barrens of New Jersey. The following morning he was dead. [2]

Colonel Joseph Bartley had the young man autopsied and sent samples of his and 19 other recruits' throat cultures to the New Jersey Department of Health. What the department found was disturbing. Private Lewis and eleven other recruits developed antibodies against samples of a virus collected from a sick pig in 1938. It looked like David Lewis had died from a swine flu similar to the swine flu that had killed 21 million people in 1918.

The Centers for Disease Control repeated the tests and came up with the same conclusion: the 1918 Swine flu had recycled back through duck and pig animal reservoirs to reemerge and kill David Lewis in 1976 - and there were three hundred more cases of flu back at Fort Dix.

Here was the dilemma. One soldier was dead and 11 other recruits were infected with what appeared to be one of the deadliest diseases known to mankind. Do you alert the world and initiate a massive vaccination program, or sit back and calculate the odds? The consequences of being wrong were so devastating that it was almost impossible to focus on the possibility that the disease that killed Private Lewis was not as contagious or as virulent as the disease that decimated the world in 1918. [3]

The director of the Centers for Disease Control, David Sencer, felt he had no choice. On a cold Saturday morning in February he convened a meeting of the nation's top medical experts. General Phillip Russell was flabbergasted when he heard the news. He was responsible for the health of the country's entire armed forces. He told the assembled doctors that they had no choice in the matter. The United States should immediately develop a swine flu vaccine and use it. David Sencer concurred, believing that since the United States had the technology and the evidence of transmission, it would be irresponsible to do anything else but develop a vaccine.

Dr. Russell Alexander expressed his doubts. Did anyone really know if the people who died in 1918 were killed from the flu alone or by *Bacilli haemophyllis*, a secondary bacterial infection that could now be controlled by antibiotics? Perhaps the government should just stockpile a supply of vaccine and hold off to see if the epidemic was really going to appear. Perhaps the flu should just be treated with Amantadine. His concerns were overridden and David Sencer had the Centers for Disease Control publish a Swine Flu Notice on February 14.[3]

But by mid-March the epidemic had not appeared. In fact the worldwide incidence of all flus had plummeted; the only cases of swine flu were still in Fort Dix, and David Lewis was still the only fatality.

But the wheels of government were already in motion. Dr. Sencer had also prepared a memo to President Ford that suggested that the government should spend $134 million dollars to vaccinate Americans against swine flu before it hit epidemic proportions in November. Gone were the uncertainties of earlier scientific reports.[4] This was the real thing, the virus that had caused the 1918 pandemic — and every American under fifty was susceptible.

President Ford did his part on national television. Flanked by Jonas Salk and Albert Sabin, the heroes of the successful polio immunization program of the 1950s, he intoned.

"I have just concluded a meeting on a subject of vast importance to all Americans. I have been advised that there is a very real possibility that unless we take counteractions there could be an epidemic of this dangerous disease next fall and winter here in the United States... Accordingly I am asking Congress, prior to the April recess, to appropriate $135 million dollars for the production of sufficient vaccine to inoculate every man woman and child in the United States."[5]

It made for great television and good politics. In the words of one congressional aide, "There was an almost unseemly race to pass the President's request". [6]

However, as summer approached, dissident scientists started to question the government's assumptions. The swine flu hadn't spread beyond Fort Dix, so was it really going to be as virulent and contagious as advertised? Perhaps David Lewis had died more from exposure than from flu alone. Perhaps those people who had died from secondary infections in 1918, could have been saved by antibiotics in 1976.

But the main stumbling block was the Pharmaceutical Manufacturer's Association whose members refused to release any swine flu vaccines unless Congress passed another bill requiring the government to assume liability. They had President Ford over a barrel. This was an election year, and he wanted to show he could do something other than pardon President Nixon.

As the summer proceeded, it looked like a stalemate. Congress wasn't going to assume liability, and the drug companies weren't going to release their vaccines without such assurances.

Everything changed when the press announced on August 2 that several veterans had died at a Legionnaire's Convention in Philadelphia. It was reported as if this were the beginning of the swine flu epidemic that everyone had been primed to expect. Scientists would later discover that it was a new bacterial illness, dubbed Legionnaire's Disease, but the damage had been done.

President Ford went back on national television to say he was "very dumbfounded" with Congress for not passing the liability bill and insinuating that Congress would bear the responsibility for millions of deaths when the epidemic arrived in the fall. Congress caved. The bill was passed. [7]

The United States government agreed to accept all liability for personal injury or death arising from the vaccination program slated to start on October 1. It also required people to sign a consent form before being vaccinated and prohibited pharmaceutical companies from making "any reasonable profit" from the swine flu vaccine. In exchange it allowed them to make reasonable profits from the Victoria strain vaccines. These provisions would come back to haunt the United States in its efforts to protect citizens against natural diseases and bioterrorism. The bill had shifted the paradigm from a realistic view of the relative risks of vaccines and epidemics to the legal fiction that the government could guarantee a risk-free world. That paradigm would soon clash with the limitations of medicine and the complexity of nature.

Meanwhile, on October 11, the UPI reported that two elderly people had died immediately after getting their flu shots. They had been given a vaccine made by Parke-Davis, the same company that, only two months before, had been forced to destroy two million doses of vaccine because it had made them for the wrong strain of flu. Confidence was slipping fast. Taxpayers had given the government $135 million dollars to produce vaccines that were now killing people. [8]

President Ford returned to the television to assure Americans that they should just ignore the scare stories and go back and get their shots, just as he and the First Lady intended to do the following morning. But the public's confidence had been broken. People stopped showing up to get their shots and doctors stopped encouraging them.

President Ford lost the election in November and David Sencer stopped the vaccination program shortly after, on December 16. In the end only 20% of the population had been vaccinated, while epidemiologists estimate that 85% of a population need to be vaccinated to provide "herd immunity" to stop a flu-like epidemic. Military strategists pointed to the dismal showing as proof that the public would fail to respond appropriately to an attack by biological weapons. [9]

And even though the program had been stopped, the cases of neurological damage kept mounting. Eventually more than 3,000 people would come down with Guillain-Barre syndrome, the illness that befell Judy Roberts. Five percent of the cases proved fatal and a quarter of the sufferers had to be put in iron lungs.

Guillain-Barre proved to be the evil doppelganger of the wildly successful polio programs that President Ford's advisors had hoped to emulate. The viruses had proved to be more clever than people. In the end, over 4,000 people would file claims against the government worth $3.2 billion dollars.[10] Judy Roberts had to plead her case on "60 Minutes" to get her share of the $93 million dollars that were finally paid out to claimants.[11]

The real dangers, however, lay in the precedents established in the hastily passed liability bill. When vaccines were first discovered people still had fresh memories of world-decimating epidemics. They were profoundly grateful that such measures could be taken and were more than happy to assume any risks involved. That was also the stance of the legal, pharmaceutical, and insurance industries. There was a realization that vaccines and humans were tricky entities. A vaccine was just a way of tricking the body into thinking it was being attacked. It was the body, not the vaccine that did the real work of conferring immunity. We also now know that not all immune systems are created equal. Older and younger people have weaker immune systems, and some people are simply better than others at fighting off diseases.

Even though vaccines had advanced from the less predictable old days of variolation, they would never be as mechanical as turning on a light switch. Antigens still had to be introduced into someone's body and the immune system had to respond. There was always the possibility that someone could contract the disease or have an adverse reaction.

Then there was the problem of contamination, the same problem that plagued Koch and Pasteur. Modern companies had to test every batch of vaccine to see that it was free of endotoxins released from the outer coating of bacteria. The old way of doing that was to inject a sample of vaccine into a live rabbit and see if the rabbit developed a fever. But in 1976 the Food and Drug Administration discovered that processed horseshoe crab blood could produce a better test. In doing so they created a new industry. Vaccine manufacturers sold off their colonies of live rabbits and started to buy horseshoe crab blood from small labs up and down the East Coast.

Today those small mom-and-pop operations have been bought out by major pharmaceutical firms, and the number of horseshoe crabs is declining. [12]

Another significant problem was public confidence. For decades after the swine flu debacle the public continued to lose confidence in vaccines and antibiotics. It would be repaired after 9/11, when Tom Brokhaw closed his evening broadcast with the salutation, "In Cipro we trust".

But most importantly of all, the swine flu nonepidemic turned the pharmaceutical industry away from developing and producing vaccines. Insurance costs were simply too steep and profits too low. Even though it was less expensive for society to prevent diseases rather than treat them, the opposite was true for pharmaceutical companies. Companies could make more money selling treatments to those who could afford them than they could selling vaccines to governments who usually could not afford them. It was this harsh fact of economics, not the dictates of better medicine that helped swing the pendulum away from public health toward insurance-covered private medicine—a story far too large to tell here.

Inevitably, these legal and business concerns would eventually come up against the laws of nature. Diseases that could have been wiped out with vaccines have bounced back stronger than ever. Bacteria have learned to evolve so fast it should make a creationist blush. By 1993 only four firms were still in the business of making vaccines. Microbes were back in the driver's seat and the world stood exposed to naturally re-emergent diseases, the growing list of iatrogenic diseases and artificially introduced biological weapons.

CHAPTER 13
The Official Story - Olde Lyme, Connecticut
1978

The official story of Lyme Disease started near the mouth of the Connecticut River and the deer filled Roger Tory Peterson National Wildlife Refuge in the leafy village of Olde Lyme Connecticut.[1]

Lyme resident Polly Murray contacted David R. Snydman of the Connecticut Health Department in 1975. Two of her children had been diagnosed with juvenile rheumatoid arthritis but she knew of several adults who had developed similar symptoms going back as far as 1968.

Dr. Snydman contacted Allen Steere, a rheumatoid arthritis specialist at nearby Yale University. They knew each other from when they had been avoiding the draft by working for the Center for Disease Control's Epidemic Intelligence Service in Atlanta Georgia.

Dr. Steere drove to Olde Lyme and interviewed 39 children and 12 adults who also suffered from what had been diagnosed as juvenile arthritis. A quarter of the patients remembered developing a strange red rash that had spread across their torsos before the onset of fever, muscle aches and extreme fatigue.

A European doctor who happened to be visiting Yale at the time told Dr. Steere that the rash looked like ones he seen in Northern Europe that had been associated with tick bites.

So Dr. Steere set about testing his patient's blood for antibodies against 38 tick borne diseases and 178 other viruses. But none of the tests came back positive and when he expanded his definition of the illness, more cases were diagnosed in adjacent states and on into the upper Midwest.

In 1977 Steere named the new disease Lyme Arthritis and published a paper about it in 1979.During his investigations Steere worked with Willy Burgdorfer from the Rocky Mountain Laboratory.

This was before the age of gene splicing and the Swiss scientist was still famous for using Swiss watch making tools and devising clever techniques to operate on ticks and mice.

One of his techniques involved strapping down mice infected with Rocky Mountain spotted fever to wooden boards surrounded by pipettes through which he force fed hundreds of ticks at a time with the blood of the pathogen filled rodents. He soon discovered that Lyme disease was caused by the Gram-negative like spirochete *Borellia burgdorferi* named in his honor.

So researchers finally had a name and a cause for the new disease. But what they didn't have was the whole story.

Everything fit together in a neat little package in what scientists often refer to as a "just so story". But the problem with "just so stories" is that they are based on beautiful yet fragile theories — theories that can be undone by just a few ugly facts.

Those facts were lurking in a dusty drawer at the MCZ Harvard's much beloved Museum of Comparative Zoology in nearby Cambridge, Massachusetts. But it would take several years to unearth them.

CHAPTER 14
The Swiss Agent
1978

Something else strange happened on the Northeast coast between 1975 and 1976. The state of New York's Health Department reported that 59 cases of Rocky Mountain spotted fever had suddenly cropped up around Long Island.

But Rocky Mountain spotted fever wasn't the only disease that emerged that summer. The first case of Babesiosis had been reported on Nantucket in 1968. But, by 1976 there were 13 cases on Nantucket, one on Martha's Vineyard and another on Shelter Island. Despite the small numbers it was the largest outbreak of babesiosis in the world.

Andy Spielman from Harvard Medical School concluded his research paper on the outbreak by saying, "We have no satisfactory explanation for the cluster of cases on Nantucket in 1975 and 1976." But another story was emerging on the other side of Long Island Sound. Little did the town fathers realize it but Lyme Connecticut would soon become infamous as the locus for the eponymously named illness, Lyme disease.

Lyme disease pushed out the other narratives and morphed into the official story, a new tick-borne disease had emerged that was hard to get but easy to cure.

Willy Burgdorfer was brought in on the spotted fever case by his former student Jorge Benach and they observed that the ticks collected by physicians and hospitals contained two types of Rickettsia-like organisms. One was clearly *Rickettsia rickettsii* the organism that causes spotted fever, and another that looked like a linked sausage but didn't react to the spotted fever tests.

But it did resemble *Rickettsia Montana* that infected meadow voles and goatherds in Switzerland so Burgdorfer nicknamed it "The Swiss Agent" and it was eventually classified as *Rickettsia Helvetica*.[1]

Later, Burgdorfer was contacted by Allen Steere from Yale's Department of Rheumatology to investigate the cases in Lyme Connecticut.

The curious thing was that the Rickettsia organism also seemed to be present in the Lyme disease cases.

So in 1978, Burgdorfer took a paid leave of absence to collect 4,000 ticks in Switzerland. When he returned to Montana he developed a new antibody test so he could rapidly detect the Swiss agent. But when he began testing the Lyme disease patients he discovered that the antibodies only reacted to the antigens on the membrane of the Swiss agent organisms. The team started to get excited, they seemed to be on the cusp of discovering a new disease organism.

But how did *Rickettsia rickettsii* get from Europe where it is endemic to the United States?

Then, in 1980 Burgdorfer made a series of phone calls to biological weapons developers on the Armed Services Epidemiological Research Board.

Afterwards he called his colleagues and told them he had run more tests that revealed some inconsistencies so they should probably concentrate on identifying the other spirochete organism in their samples.

Eighteen months later the team announced that Lyme disease was caused by *Borellia Burgdorferi*. No mention was made of Rickettsia or co-infections that could make Lyme disease more difficult to identify and far more difficult to cure.

This myopic focus on Lyme disease to the exclusion of Rickettsia and other co-infections has led to the controversy over so-called chronic Lyme disease.

So today about 7,000 people a week contract Lyme disease and about a third of them develop severe joint, heart and neurological problems sometimes leading to death.

Their numbers far outweigh the number of AIDS and Breast Cancer patients combined but only a fraction of the amount of research funding is going into Lyme disease and almost nothing is being done to screen for the Rickettsia Swiss agent.

But how did the Swiss agent, endemic in Europe, get into American ticks? And there was another twist to the story. Lyme disease and Rickettsia were both emergent and expanding throughout Russia. How did the disease organism get there?

CHAPTER 15
Outbreak - Plum Island
1978

Michael Christopher Carroll visited Plum Island several times to research his book *Lab 257*.

On his last trip he and Ben Robbins, a former draftsman on the island, beat their way through overgrown brush to climb to the top of the high bluff where the Army had first taken measurements to see if Plum Island was a safe spot to release pathogens.

After the tour Michael looked down at his khakis and a flash of heat surged from his toes to his gut. Eight tiny poppyseed sized deer ticks were clinging to each of his pants legs. When he finally had a chance to strip down in the men's room he discovered nine more ticks attached to his shirt and inside his pants. He had been to many hotspots and never encountered so many ticks. Earlier their van had filled with a sickening stench and they had seen a thick cloud of smoke rolling toward Gardiner's Island, Sag Harbor and the ever-popular Hamptons.

Bob twisted around in his seat, saying, "Time to burn the animals," referring to the livestock killed by exposure to diseases like hoof and mouth, rinderpest and Rift Valley fever so closely related to the Ebola virus.

But Michael had seen too much. He was not allowed further access to the island.

The Plum Island Animal Disease Center had been started in 1951 and run for 25 years by the U.S. Department of Agriculture with the U.S. Army as a silent partner. The Army initially wanted to study using livestock diseases to destroy the food supply of the Soviet Union as a way to weaken the Soviet government and encourage defection. Later they became involved in more traditional biological warfare.

The Center's first administrator was "Doc" Shahan who made his name by using 10,000 soldiers to contain hoof and mouth disease before it entered the United States from Mexico.

In its early days the center was considered to be 'the safest lab in the world," and was know for its flowerbeds and well-kept grounds.

In 1981 Dr. Endris came to Plum Island to expand the Center's tick research. He built two insectaries, one in the basement and one in the back of Lab.[1]

He raised over 200,000 hard and soft ticks of multiple species that he dosed with disease organisms by letting them feed on the bodies of infected pigs, goats, mice and calves. Endris had worked on earlier tick research on the island even though a fellow worker said that Plum Island, "wasn't set up to deal with ticks at all." It was during this time that Lyme disease and Rickettsia and Babeosis suddenly appeared on Martha's Vineyard, Nantucket, Long Island and in Lyme Connecticut.[1]

Later, Michael asked Willy Burgdorfer whether he thought there was a connection between research on ticks on the island and the appearance of Lyme disease in Lyme Connecticut.

"Touching on that would raise a hell of a lot of problems. Unless they cultivated the ticks species on Plum Island and unknowingly fed some of the ticks on animals or humans and a Borellia spirochete accrued."

"Plum Island is proof of the existence of breaks in biological safety. Even if a lab has the highest containment level that doesn't mean it is safe."[2]

But that is not the only way ticks could radiate out from Plum Island. Birds are notoriously "ticky" and Plum Island lies in the middle of the Atlantic flyway for over 140 species of resident and migratory birds.

Canada geese, osprey and seagulls nest on the island and make daily trips to Connecticut and Long Island. Golden and bald eagles prey on the massive flocks of migratory Canada geese that local birders refer to as the Canadian Air Force.

In the spring thousands of barn, bank and tree swallows gather on Plum Island to rest before migrating to Maine and Nova Scotia.

They wait until they build up a critical mass before pushing north, or south in the fall. Their first landfall is Long Island going south and Connecticut going north.

Then there are the deer that swim back and forth from the island to Connecticut and Long Island's Orient Point. Sometimes a bloated dead goat or sheep will drift over to Connecticut or Montauk and reportedly be retrieved by men wearing pure white biohazard suits. Then there is the line from, "Silence of the Lambs" where Clarice tells Hannibal Lector she has found him the perfect place for a summer vacation... the beaches of New York's Plum Island. He declines the invitation to "Anthrax Island."

But back to reality. In 1978 one of the technicians entered the lab's livestock cubicle and was horrified to see two steer stumbling around their cell-block groaning, drooling and foaming at the mouth.

The animals should have been clean. The animal handlers had brought them in from their outdoor pens only the day before. He called Dr. Dardinni who had just disembarked from the morning ferry.

After examining the steer Dardinni confirmed their worst fears. The steer had the unmistakable signs of the lab's nemesis hoof and mouth disease, as did all the other cattle in the Old Cow Barn.

The outbreak called for what the Center's emergency manual referred to as Armageddon, the destruction of all the livestock on the island; in this case 94 cattle, 87 pigs, 66 lambs, 28 rabbits, 27 chickens, 13 goats, 6 horses and small colonies of mice and guinea pigs.

Workers on the island were ordered to stop work and await further instructions. The director radioed the Plum Islam ferry and ordered the Captain to turn around so contractors would not spread the virus to the mainland. It wasn't until nightfall that the workers were decontaminated and allowed to return home wearing baggy white spacesuits.

After the contractors left, animal handlers herded steer, goats, sheep and horses out of their pens and into the lab. On Saturday morning they fired up the incinerators and led each cow down the disassembly line where Dr. Dardinni drew blood and clipped off bloody samples of body tissue.

Then two handlers used anesthetics and an air compressor gun to shoot bolts into the heads of the cows before other workers used power saws and large knives to reduce the cows into big chunks of flesh, which they shoved down the chute into the roaring incinerator.

It was a macabre sight seeing the detached heads of cattle, sheep and horses roll down the bloody chute after their legs midsections and entrails. All to the sounds of saws, chains and the bellows and squeals of the terrified animals.

By the end of "kill weekend", the necropsy room was strewn with blood and pieces of flesh and the bricks of the incinerator ran so hot that its 84-foot chimney glowed in the night.

CHAPTER 16
Sverdlovsk
1979

There is a small group of analysts and scientists within the American government who have always felt the United States made the wrong decision when it stopped making biological weapons in 1969. Some went back to Fort Detrick; others were analysts in the CIA and the State department. They were probably aware that the CIA continued to stock agents in violation of the biological weapons treaty and probably supported American research that used gene splicing techniques to insert plague-like microbes into E.coli in order to increase its "level of pathogenicity". [1]

These people could justify such research because they believed the Soviet Union was still amassing biological weapons, and they interpreted the treaty to mean that nations could perform such research for defensive purposes. In other words it was acceptable to design superbugs as potential weapons in order to find ways to defend against them. The problem was that the American government provided money for at least 51 projects to make novel pathogens but nothing for research to defend against them. [2]

Immediately after the biological weapons treaty went into effect, this small band of hardliners started to meet at monthly brown bag lunches to share information about possible violations. They were informally headed by Gary Crocker, an analyst in the State Department, and became known as the "bugs and gas" guys. The bugs and gas boys had access to CIA intelligence but were often thwarted by their superiors who didn't want to believe them or softliners who didn't want to rock the boat with the Soviet Union.

Softliners felt they had bigger fish to fry, like the reduction of strategic missiles and the destruction of nuclear weapons.[3] However, it looked like the hardliners would finally get their break in 1979. Instead, a media-savvy civilian scientist would deny them their smoking gun for over a decade. It all started near the Ural Mountains in March 1979.

....

Lieutenant Colonel Nikolei Chernyshov walked slowly down the tree-lined streets of Sverdlosk. The apparatchiki in Moscow had done their best to erase the pre-Revolutionary history of the grand old capital of the Urals. Gone were the days when Yekaterinburg used to supply emeralds and malachite to the Romanovs. The house on the bluff overlooking the Iset River was a stark reminder of how much had changed. It was where Lenin had the Tsar's family held before they were brutally murdered in 1918.[4]

First, the apparatchik had renamed the city after one of their own, Yakov Sverdlosk, the first president of the Soviet Union who had died of the Spanish flu in 1919. Then they had slowly, yet inexorably turned the old mining town into a military-industrial complex closed to the outside world.

Colonel Chernyshov worked in the heart of that complex. He slowed as he approached the heavy wrought iron gate of Compound 19. The ornate patterns of the electronically controlled gate were all that reminded Chemyshov of the old days of Yekaterinburg. The rest of the building was purely functional. It had been designed using the captured blueprints of General Ishi's biological weapons complex in Ping Fan. Two thousand troops now worked and lived in the military garrison. It was the Soviet Union's largest manufacturing facility for weaponized anthrax.[5]

Chernyshov was head of the night shift on Compound 19's main production line. His workers were in charge of fermenting, drying, milling, and aerosolizing the weapons-grade anthrax spores.

Russia had an age-old relationship with anthrax. Ilya Metchnikoff had supplied the Tsars with the their own private supply of anthrax vaccine when he worked for Dr. Pasteur in 1888. The serfs used to call anthrax "Siberskaya yazva", Siberian ulcer, for the horrible lesions it caused in people who handled infected wool.[6] Tonight, Colonel Chernshov was going to make his own batch of Siberian ulcer.

Cherynyshov's seed stock came from a rat discovered in the sewers of Kirov. In 1953 a leak in the Kirov bacteriological facility had released some anthrax into the local sewer system where it mutated in the rodent population. Three years later a Soviet scientist captured a rat infected with the new, now highly virulent, form of anthrax. The Soviet army named the new strain Anthrax 836 and made it their weapon of choice against the West.[7]

Chernyshov's team was in charge of drying the slurry of liquefied spores after they had been sluiced out of the story-high fermenters. Before the spores could be milled into their small aerosol form, they had to be run through the compound's massive dryers. An exhaust system blew the excess heat through a 24-inch vent at the top of the building.

By the end of the shift Chemyshov and his workers were tired and ready to go home for the weekend. Most would head straight for the nearest bar. Chernyshov would go home to his family. This weekend they planned to visit some of the ornate wooden cottages that still lined the outskirts of the city.

Just before the shift ended, Chernyshov gave the signal to shut down the heaters. As the roar of the blowers subsided, technicians scrambled over the machinery to inspect it for Monday's early morning shift. The workers hated this time. It was when the spores were most likely to float up off the dryers. Even though they had all been vaccinated, nobody liked to think that they could be inhaling great lungfuls of Siberian ulcer.

One of the technicians climbed to the ceiling to inspect the 24-inch filter that was all that separated the exhaust heat from downtown Sverdlosk. A thick paste of muddy gray anthrax spores clogged one of the screens. The technician unclamped the filter and scribbled a note to Chernyshov.

"Filter clogged so I've removed it. Replacement necessary."[8]

Later Colonel Cherynshov could not remember why he had failed to copy the note into the logbook. Perhaps he was thinking of the upcoming weekend, perhaps he was overtired from work. Whatever it was, he would have to live with this momentary bout of forgetfulness for the rest of his life.

Early Monday morning, the dayshift manager scanned the logbook. Seeing nothing wrong he ordered his technicians to start up the dryers. A fine mist of anthrax spores shot out of the exhaust vent into the pre-dawn air. For several hours the spores drifted silently on the cool northwest wind. Some were inhaled by reservists in the adjacent compound 32, others by pipefitters in the ceramics plant two kilometers away, and more floated to villages 80 miles downwind.

As the dormant spores became active people started collapsing in streetcars and dying in the lobbies of buildings. Spores were binding to the victim's phagocytes cells. They were forming hetamers, syringe-like organelles that injected virulent factors into the victims' immune systems. Edema factors released by the spores caused tissues to swell and lethal factors caused organs to collapse and die. [9]

Doctors did not know what they were dealing with. Victims were staggering into hospitals running 100-degree temperatures and drowning in their own fluid-filled lungs. Soon the two local hospitals were full. Doctors gave patients every antibiotic they had on hand, but it was too late; the toxins had already started their action. Nobody told the civilian doctors that they had been practicing next to an anthrax factory. They assumed they were dealing with a highly infectious new form of pneumonia of unknown etiology.

After the first week, orders came from Moscow to quarantine the area and transfer all patients to Hospital 40 for screening and autopsies.

By this time some people had started to suspect that compound 19 could be the source of the infections but its work had kept so secret that even the local KBG chief did not know what had been going on in the facility. However he did have initiative and phone taps.

Soon he knew of Colonel Chernyshov, the missing filter and the inadvertent release of anthrax.[10] But this was something that could never be admitted to the public or to the West.

Now the KGB in Moscow took charge. KGB officers posing as doctors visited the homes of the deceased and persuaded family members to sign falsified papers. They seized hospital records, changed dates, and deleted all reference to inhalation anthrax in the autopsy reports. Residents were told that the deaths were caused by a truckload of tainted meat. Hundreds of dogs were rounded up and killed, and several black market vendors were thrown in jail. It was all part of a well-crafted cover-up that claimed that people had died of intestinal anthrax from eating tainted meat, not from inhalation anthrax from biological weapons.[11]

The Communist party prolonged the epidemic by ordering the city workers to scrub down roofs and spray trees. This only re-suspended the spores and led to a second wave of inhalation and some intestinal anthrax cases. Eventually 66 dead victims were drenched in chlorinated lime and secretly buried in the isolated Rostochniy cemetery. [12]

A few months later, Kanetyjan Alibekov was furious when he discovered that one of his employees at the Stepnogorsk plant in Kazakhstan was Colonel Chernyshov. Nobody had bothered to tell him that this man, who was his expert in anthrax drying techniques, had been responsible for killing 66 innocent people. Dr. Alibekov pitied the broken man before him but was angered that

Chernyshov had never been punished and was still free to make biological weapons. But Alibekov's hands were tied. Evidently Chernyshov still had the blessing of the KGB that had the final say on all hiring decisions.[13]

The West heard about the incident when an anti-Soviet newspaper published an account in Frankfurt, Germany. The article claimed that an explosion at a military base had led to the release of a cloud of deadly bacteria that had killed a thousand people. The official news agency, Tass, denied the allegations quoting officials who stated that the epidemic had been caused by a natural outbreak of anthrax among domestic animals.

Shortly after news of the incident hit the American press, the CIA flew Matthew Meselson to Virginia to show him their satellite photos of Sverdlosk. For several days he stayed at a CIA staffer's private home.

"I got hooked on this thing and became somewhat skeptical of the CIA's explanation," said the eminent biologist.[14]

It had been Dr. Meselson who had convinced the Nixon Administration to stop making biological weapons, and Dr. Meselson who had become the treaty's most ardent supporter. But Dr. Meselson had another claim to fame. During the Reagan Administration he had become famous for debunking the CIA's belief in "Yellow Rain."

In early 1979 Hmong tribesmen reported that helicopters flown by Soviet-sponsored troops had sprayed them with a mysterious yellow substance that burned their skin and left telltale yellow spots on vegetation. The mysterious agent soon became known as "Yellow Rain," suspected by the bugs and gas crowd to be a biological weapon made from fungus toxins.[15] Dr. Meselson belittled their theories in a series of articles that presented evidence that showed that the yellow stains probably came from bee feces rather than from biological weapons.

The articles in Foreign Affairs and Science Magazine made the Reagan Administration look particularly foolish. Undoubtedly, that triumph had primed Dr. Meselson to take a skeptical look at the government's explanation of Sverdlosk as well. Skepticism is a prerequisite for science and can keep a democracy healthy and honest, but in this instance it hid the truth for over a decade.

By 1986 Dr. Meselson was convinced the bugs and gas guys were wrong about Sverdlosk as well. He flew to Moscow to meet with Soviet health officials who had been present at the time of the accident. Dr. Pyotr Burgasov convinced him that the people who died in Sverdlosk had contracted intestinal anthrax from eating cattle that had been fed bone meal from a contaminated bone meal plant. It was all part of the Soviet Union's well crafted cover story.

But the disinformation campaign worked better than the KGB could have ever hoped for. Dr. Meselson was an influential member of two of America's most prestigious intellectual organizations, the National Academy of Sciences and the American Academy of Arts and Sciences. He was also an old friend of Alexander Langmuir, head of the Johns Hopkins School of Public Health. Dr. Meselson arranged to have Dr. Burgasov and two colleagues present their evidence at all three institutions.

The lectures became the must-attend events of 1988. They drew the intellectual elites of Cambridge, Baltimore, and Washington. Dr. Burgasov assured his prestigious audiences that there was no doubt the deaths were caused by food poisoning and that a military accident was out of the question. Dr. Meselson added his imprimatur by saying "the evidence was plausible and consistent with what was known from the medical literature and recorded human experience."[16] The bugs and gas guys had been trumped again.

Back in the USSR, Kanatyjan Alibekov was amazed that the West had actually fallen for the cover up. He had refused to sign off on Dr. Burgasov's papers thinking they would only make the Soviet Union look foolish.[17]

That was where things stood until Mikhail Gorbachev ushered in the era of Glasnost. In 1990 Soviet journalists gradually started to uncover what had really taken place at Sverdlosk by interviewing some of the doctors who had treated patients in hospitals 20, 24, and 40. Their stories were picked up and repeated in the West.

Then in 1991 the Wall Street Journal sent Peter Gumbel to Sverdlosk to talk to doctors and family members of the deceased. Among other things, he made the telling discovery that the bone meal plant that had supposedly supplied tainted cattle food simply did not exist. [18]

Russian papers also reported how high up in the Soviet hierarchy the cover-up had reached. Yuri Andropov had ordered the cleanup of Compound 19, and

Boris Yeltsin became furious when he learned that he had been left in the dark during the crisis when he had been Sverdlosk's Communist party boss.[19]

Even Dr. Meselson eventually became convinced. In 1992 he, his wife, and several colleagues were given permission to visit Sverdlosk and interview family members of the deceased. It was an extraordinary piece of alchemy that after September 11 their book *Anthrax* was hailed as the definitive book on Sverdlosk, despite the fact that its authors had been duped for over a decade. The bugs and gas guys had been correct all along, but they would have little time to gloat. They would soon have their hands full with other, more pressing, concerns.

CHAPTER 17
The Phone Call - Biopreparat
October 30th 1989

It was October 30, 1989. A light rain drizzled on the almost empty streets of Moscow. A government-owned black Volga prowled silently through the gray leafless northern section of the city. It paused at the sidewalk, its motor still purring. Colonel Katanya Alibekov descended from his well-appointed apartment. He gave his driver a warm greeting, then feigned surprise.

"No bodyguards today?"

"Not until your meeting at Army headquarters, sir."

"Ah, good." [1]

Savva smiled but did not respond. It was part of the unspoken understanding between the two men. Neither the Deputy Director of Biopreparat nor his driver liked the sullen presence of the bodyguards who were required to be in the car when secret documents were being transported. Colonel Alibekov was never sure whether the KGB agents were there more to defend him, or the papers.

Normally loquacious, Alibekov seems preoccupied. It was that damn meeting coming up in Protvino. The heads of the thirty-some institutes under his command would be there. All of them were behind schedule and he needed to prepare a stern rebuke. No time to ask after Savva's wife and family. No time to swap stories of their children's' escapades. As Alibekov put the finishing touches on his impromptu remarks, Savva glided the limousine through the early morning traffic, occasionally flashing the Volga's official blue lights to warn policemen that an important personage was passing through.

Colonel Alibekov had come a long way since his early days as an honors student at the Tomsk Medical Institute — the car, the apartment, the good schools, the triple salary as a bureaucrat, medical doctor, and military officer. He realized how much he and his family had grown accustomed to the many perks of being in the upper echelons of the Soviet Union's military industrial hierarchy. How different it was from the life he had envisioned for himself as an idealistic young doctor from Kazakhstan.[2]

Colonel Alibekov's early work with tularemia had earned him the attention of General Kalinin, the powerful head of Biopreparat, the Soviet Union's government-run pharmaceutical empire, created after the Soviet Union signed the Biological Weapons Treaty in 1972. It had been a brilliant maneuver to hide the Soviet Union's ongoing biological warfare program within a so called civilian agency. Yet the move had created bureaucratic friction between the upstart new agency and the older military programs. General Kalinin had instinctively known that Alibekov's success at weaponizing tularemia had presented him with a winning hand in his ongoing battles with the Fifteenth Directorate. [3]

In 1983 Kalinin had offered Alibekov the job of upgrading the Stepnagorsk facility to replace Sverdlosk. Military officials bristled at the effrontery of hiring someone at the unheard of age of 32 to oversee a project almost certain to fail. But it was an ideal situation for the crafty general. If Alibekov succeeded Kalinin would get all the credit, if the plan failed Alibekov could be blamed. One of Alibekov's friends expressed it best: "Now you are the age of Christ. It's a good age to die a martyr." [4]

But four years later Stepnagorsk was producing two tons of anthrax a day, Alibekov had made the institution the most efficient anthrax production facility in the world and cemented his reputation as someone who made things happen. Even KGB officials had to respect Colonel Alibekov when he bent their security rules to hire the scientists he needed.

But it was still a shock when General Kalinin asked Colonel Alibekov to come to Moscow to be his second in command at Biopreparat. Mikhail Gorbachev had ordered Biopreparat to start research on the production of biological weapons from such exotic viral diseases as smallpox, Ebola, Lassa, Machupo, and Marburg fevers.[5] General Kalinin knew his young protege could get the job done.

By 1992 Colonel Alibekov had succeeded again. Biopreparat's far-flung empire of research and production facilities were loading a hundred tons of smallpox a year into ICBM missiles aimed at over a hundred Western cities including Washington, New York, London, Chicago, and Los Angeles. The military had modified its cruise missiles so they could fly below enemy radar to lay a swath of deadly viral agents onto cities, troops, and civilians. The stated purpose was to ensure the West's destruction in the event of an all-out nuclear exchange.[6]

....

Savva pulled into the gated courtyard of Pyotr Smirnoff's yellow brick mansion on the tree-lined side street of Moscow's old German Quarter. The vodka merchant's former estate on Smokatnaya Street was the now the discreet new headquarters of Biopreparat.[7] Colonel Alibekov climbed the broad marble stairs and entered his office where a thick pile of progress reports awaited his perusal.

But his secretary handed him a more urgent telephone call. It was from the deputy director of the Institute for Ultrapure Biopreparations. Alibekov picked up one of the five phones that connected him to the Kremlin, his institutes, the KGB and the military. The news was not good. The director of the Institute, Vladimir Pasechnik, had not returned from Paris.[8]

Colonel Alibekov sighed and picked up the phone that connected him to Savva Yermoshin, his KGB chief and old friend from their early days at Stepnagorsk.

"Savva, we have a big problem."

"You always have big problems over there Kanatjan."

"No Savva this is real. I think Pasechnik has defected." There was a long pause on the other end of the phone.

"Shit!" [9]

"Easy for you to say," thought Alibekov to himself, "I was the one who gave him permission to go to Paris." But Kanatjan was never blamed for Pasechnik's defection.

A few months later General Kalinin asked Colonel Alibekov to step into his office. Moscow had received a highly unusual diplomatic demarche. It stated that the United States and Great Britain had "new information" that the Soviet Union was violating the Biological Weapons Treaty. Alibekov had never seen General Kalinin so upset. He knew the "new information" had come from Pasechnik.

"We're going to have headaches from now on," warned Kalinin. "Shevardnadze is furious. They say he stormed into Gorbachev's office to demand what was going on. Apparently our foreign minister doesn't like to learn what's going on in his own government by foreigners." [10]

Pasechnik had not only told British intelligence the details of his research at the Institute for Ultrapure Biopreparations, but he had also revealed that the real purpose of Biopreparat was to act as a civilian cover to allow the Soviet Union to continue making biological weapons in violation of the Biological Weapons Treaty. The cover had reigned for twenty years as the Cold War's best-kept secret.

Even the West's most hawkish analysts had always assumed the military was in charge of the Soviet Union's biological weapons. Nobody had guessed that the program had been quietly lurking under the cover of Biopreparat. It is difficult to convey how shocked the CIA was when it was informed of Pasechnik's revelations. The Soviet Union had become the world's only biological superpower without the West's knowledge. It was the Cold War's most spectacular intelligence failure - one that continues to reverberate through today's headlines.

At the time, the demarche was nothing but a huge distraction. It had to be answered. Colonel Alibekov was forced to sign off on a document that insisted that the USSR had only conducted research for defensive purposes. But the reply added an interesting twist. The Soviet Union agreed to open its facilities to Western inspectors. It was a ploy, of course. Colonel Alibekov never thought the United States would agree to inspections. It would force the Americans to reveal their own program, which they had, of course continued to pursue after signing the biological weapons treaty. [11]

A month later Colonel Alibekov received a rude awakening. After a series of meetings with top army and KGB intelligence officials, he realized they could tell him nothing about the American biological weapons program. Was it possible that the United States really did not have such a program?

His entire life had been built on the premise that the Soviet Union had to build biological weapons because the United States was building them as well. If the United States was not building biological weapons what had he done with his life? Why had he taken the Hippocratic oath "to do no harm"? Alibekov was so shaken by the realization he had to excuse himself from the KGB meeting in order to compose himself. [12]

By 1990 Alibekov did not have much time to think about the American program. Too much was changing in the Soviet Union and in his own life. The military saw Mikhail Gorbachev's call for glasnost and perestroika as a way of ridding themselves of their old nemesis Biopreparat. In a heated argument, Alibekov had tried to convince General Kalinin that the best way to save their jobs was to convince Gorbachev that Biopreparat should stop making biological weapons and concentrate on legitimate pharmaceutical and biodefense research. To his astonishment not only Kalinin but the KGB had finally agreed. Perhaps the KGB knew all along that the United States had not really been making biological weapons.

By this point Alibekov had already decided to separate himself as much as possible from his deputy director's duties at Biopreparat by becoming the part-time head of Biomash, a scientific design unit that provided technical equipment to hospitals, medical labs, and weapons facilities. It was at least a step in the right direction. But on January 11 1991 Kalinin called Alibekov into his office to ask a special favor.

"Kanatyan, the American and British delegation is arriving on Monday. I'm too busy to show them around would you mind acting as host?"

Colonel Alibekov did mind. He had just started his job at Biomash. Plus he did not see how he, or anyone else, could possibly prevent the Americans from learning what the Soviets had produced. Kalinin assured Alibekov that all the institutes had been sanitized so they looked like they had only been producing pesticides and vaccines.

"Kanatyan, we need your help!"

Colonel Alibekov reluctantly agreed. [12]

The inspections did not get off to an auspicious start. The windshield of the bus Biopreparat chartered to show the inspectors around Moscow exploded in the frigid morning air and the inspectors and their handlers ended up huddled for warmth in a massive group hug in the back of the bus. It was the first time Alibekov had ever seen, let alone hugged, an American. [13]

But Colonel Alibekov received his first ugly surprise at the Obolensk Institute where Biopreparat scientists had been using gene-splicing techniques to develop drug resistant strains of anthrax, plague, tularemia, and glanders, the disease that German soldiers had used against Russian horses during World War One. [14]

General Urakov, the director of the Institute, asked the delegation what they would like to see. Chris Davis, the head of the British team pulled out a map and pointed directly at Obolensk's massive explosion chamber. Alibekov cursed to himself. The map must have come from satellite photos. A few moments later Davis was at the closed door asking what was inside.

"We lost the key. I'll see if I can find a copy," stalled one of the senior scientists.

"Can you turn the lights on?"

"The bulb is out."

Davis pushed past the scientist and pulled out a flashlight. Chief scientist Petrukhov grabbed Davis' wrist and the two grown men grappled in the hall. Scientists separated the two and Alibekov asked Davis what was wrong.

"This KGB guy tried to grab my flashlight."

"Really, that man is a respected scientist. He is not a member of the KGB." But Alibekov knew there was nothing he could do. Flashlights were not forbidden by the rules negotiated for the visit.

Davis went into the room and saw where the walls and door were severely dented. It was obviously from explosives.

"No, no," said Petrukhov. "Those marks come from the hammers we had to use to make the door fit."[15] Alibekov winced.

The inspection at the Vektor Institute went no better. First the plane carrying the inspectors to Siberia was forced to spend the night in Sverdlosk, of all places. Inside the Institute Alibekov watched the Westerners' eyes widen as he took them past huge fermenters larger than anything necessary for making vaccines.[16] He did manage to prevent them from taking smear samples that might have revealed the presence of gene-altered smallpox, Ebola or any of the other exotic hemorrhagic fever viruses studied at the facility; nor did they see Koltsovo's nearby Building 15, which was capable of producing a hundred tons of smallpox a year.

Things got worse when they visited Pasechnik's old lab, the Institute for Ultra-Pure Biopreparations in Leningrad. The new director made a point of not mentioning Pasechnik's name but the ever-pesky Davis found an old bulletin board with an announcement signed by the former director. Later, Alibekov raged silently at the sanitizers when the visitors stopped in front of Pasechnik's old machine for jet milling anthrax. No one had told him that the incriminating machinery had not been removed.

"What's this?"

"For salt," chimed in the institute's deputy director. "That's where we mill salt for medicinal purposes." [17]

Even Alibekov could not suppress a smile at the ridiculous, albeit quick-witted response.

Alibekov was relieved when the visit was finally over. He had enjoyed meeting his fellow scientists and had felt stupid trying to deceive them with lies he knew they had not believed. At the final banquet he offered a heartfelt toast. He apologized to the British and American visitors if they thought the Soviets had not been very open, admitting that they had their secrets in biodefense, but after all this would not be the last visit by the British and Americans. He said that that he looked forward to being their guests soon. He closed by saying that a lot of Soviet people supported the American actions in Iraq, adding, "I truly hope you win."

After dinner Savva Yermoshin led Alibekov into a quiet corner, "Kanatyan, I think you should stay away from politics." [18]

The final report of the joint team of inspectors was 200 pages long. It detailed their evidence that Biopreparat had been running the world's largest biological weapons program, and it reiterated that they had not even been invited to visit the military's production facilities. But the report was kept secret as had been agreed upon by President Bush and Mikhail Gorbachev. Besides, the world had other concerns: the United States was battling Saddam Hussein and the Soviet Union was preparing to implode.

CHAPTER 18
Defection
1992

"This is awful," muttered Grigory Berdennikov as he unpacked his bags after the long flight to Washington. The American newsman was announcing that the leaders of Russia, Belarus and the Ukraine had just agreed to form the Commonwealth of independent States. Colonel Alibekov agreed, there was no hope for Gorbachev now.

"You don't understand, we're carrying passports from an extinct country. The Americans will probably tell us to go home." [1]

It was not an auspicious start to the Soviet Union's reciprocal visit to America's biodefense establishments. There were already divisions within the team. Berdennikov represented the foreign ministry. He had been a close ally of Edvard Shevardnadze and was secretly encouraging the US State department to find out more about the Soviet biological warfare program. Neither Gorbachev nor Yeltsin believed they were being told the full truth by their military commanders or each other. [2]

Colonel Alibekov was head of the scientific team along with Dr. Lev Sandakhchiev, director of Vektor, and General Nikolai Urakov director of Obolensk. Alibekov had little use for Dr. Urakov, who had sided with the illegal putsch that had almost overthrown Gorbachev in August. Colonel Vasiliev represented the military who had their orders from General Yevstigneev to return to Moscow with proof of American offensive biological weapons - no matter what they actually found.

The Soviet team flew to the Dugway proving grounds aboard Air Force Two, the Vice President's luxurious hundred-passenger jet. What a delightful contrast to the flight to Siberia, mused Alibekov as he sampled the food and wine.

But it was Lisa Bronson who captured Alibekov's full attention. The attractive young woman led the American team, and even top military officers seemed to scurry to fulfill her requests for access, information, and transport.[3] Alibekov liked her immediately.

Chris Davis and David Kelly greeted Alibekov warmly. The British experts remembered him from their first visit to the Soviet Union and were amused that he was no longer wearing the old brown cardigan that had so mystified them on their former trip. They had assumed he had been wearing it to conceal "some sort of secret equipment".[4] Little did they know that Alibekov had been wearing the old sweater for several months as a form of protest against the rigidity of General Kalinin and Biopreparat.

The Americans provided helicopters to ferry their visitors to whatever far-flung labs, sheds, and former testing facilities the Soviets wished to see. But Colonel Alibekov could see no evidence that the Americans had been recently testing weapons. Where were the animals, cages, and test equipment? The doors of the antiquated buildings were swinging on old rusty hinges and paint was flaking off the walls. Assistants in the laboratories answered the visitors' questions fully, explaining that they were using bacterial simulants like Serratia to explore ways of protecting soldiers from biological weapons. It was clearly defensive work as allowed under the biological weapons treaty. Toward the end of the visit, Lev Sandkhachiev sidled over to Alibekov.

"They're doing nothing here." [5] Colonel Vasiliev was visibly annoyed. He poked through the scrubby soil to collect samples of soil, grass, and a fragment of debris that he would later claim was part of a bomblet. Dr. Urakov said nothing.

When they arrived at Fort Detrick a few days later Colonel Alibekov was more reassured. This looked more like a testing facility. It even had a million-liter explosion chamber. But, "The Eightball" had not been used since 1969, Lisa Bronson explained, pointing to the brass plate that declared it a national monument.

Back on the bus, Colonel Vasiliev pulled out a map. "What's this building?" he asked pointing toward a circular shaped structure that his military briefers in Moscow had assured him was a weapons testing facility. The American guide whom somebody had nicknamed "Smiley" seemed confused. He went to the back of the bus to confer with the other members of the US team. Finally he returned with the assurance that there was nothing there. Alibekov smiled. Did the Americans take them for fools?

After driving around for twenty more minutes Dr. Bailey finally located the structure that rose like an inverted ice cream cone. Huge bay doors stood ajar and a pile of ominous gray powder lay on the ground. The Soviets conferred amongst themselves in rapid fire Russian. The interpreter asked what it was. Smiley responded that it was salt, "We use it to cover the roads in winter."

Alibekov guffawed. It was the same response they had used to explain Paschnik's milling machine in Leningrad. Vasiliev was angry. He strode toward the pile, thrust his finger into the powder and jammed it into his mouth. The Soviets waited impatiently for his response.

"It's salt." [6]

Alibekov was becoming amused at the absurdity of the situation. It was clear that nothing had been going on at Fort Detrick as well as at Dugway. The only thing that had bothered him was "Smiley's" demeanor. Whenever he smiled Alibekov scowled back at him. The deputy director of the Army's Institute of infectious Diseases had to be hiding something. Charlie Bailey had just assumed that Alibekov was scowling because he was really a spy.

Things became more surreal at the Pine Bluff Arsenal. This was the Arkansas home of the infamous "Pine Bluff Cocktails" designed to be dropped on Cuba — the Soviet Union's staunchest ally. Officials at the facility assured the Russians that the installation had been turned over to the FDA in 1969. Unfortunately for the Soviet delegation, the evidence seemed to support the claim.

Colonel Vasiliev asked to see the vats where entomologists used to breed swarms of mosquitoes raised to act as vectors for biological agents. The vats were now full of catfish, being grown for biomedical research. While the Russians were being treated to a long dissertation on catfish biology, Colonel Vasiliev videotaped the valves and pumps. The plumbing was the newest equipment the Russians had seen on their visit. In Moscow, Vasiliev would use the videotape to show that the catfish were just a cover-up to camouflage the true nature of the dual-use facility.

Colonel Alibekov discovered a notebook locked away in a contaminated waste treatment building. He blew the dust off the notebook and scanned the handwritten notes. They had been written in 1973. There was a centimeter of dust on the floor, and the insulation wrapped around the tanks was cracked and wrinkled with age. It was obvious the building had not been used for decades.

Next was Pine Bluff former facility for filling bomblets with biological agents. It had been converted into research space. A dozen scientists were now busily grafting pieces of bird tissue onto the backs of mice to study immunosuppressive substances. Lev Sandakchiev could not stop asking questions. The scientists on the Soviet team had grown bored of the search to find biological weapons when it was obvious there were none to be found. They queried their scientific colleagues for hours, much to the annoyance of their military counterparts, who regarded such idle chitchat as fraternizing with the enemy.

But on the following day, the military had their turn. Alibekov was dozing when suddenly Colonel Zukov jumped out of his seat demanding that the bus be stopped.

"We have to check that out," said Zukov gesticulating toward a towering structure of metal.

"Don't be ridiculous. It's a water tower."

"I don't think so."

Zukov jumped down off the bus and proceeded to climb the tower. One of the Russians snapped a photo as he approached the top. The Americans stifled their smiles. When Zukov finally returned to the bus a colleague asked what he had found.

"Water," he snapped.[7]

The last stop of the tour was the Salk center at Swiftwater, Pennsylvania. In the 1950s Swiftwater scientists had produced smallpox vaccine from cowpox, a supply of it was still held in cold storage. The Soviets were never shown the vaccine. It would have made them suspicious.

But Colonel Alibekov remembered their final day in Washington most clearly. Lev Sandachiev asked Lisa Bronson how much money a scientist could actually make in the United States. Bronson replied that it depended on your experience. A government scientist could make between $50,000 and $70,000 dollars, but a scientist in the private sector can earn up to $200,000 a year.

Someone released a low whistle. Alibekov asked, "With my experience, could I find a job here?"

"If you knew English," was Lisa's retort.[8]

Privately Alibekov thanked Lisa for her help and requested a business card. She was pleased, at least Colonel Alibekov and Grigory Berdennikov seemed to understand the US really didn't have anything to hide. She only wished she had convinced the military members of the Russian team.

Colonel Alibekov returned to a country that was no longer his own. Russia's tricolor flag now hung over the Kremlin where the hammer and sickle had hung before. As a Kazakh, should he go home to Kazakhstan or continue to live as an alien in Russia?

In January of 1993 Alibekov resigned from the army and Biopreparat. He quickly landed a job as the Moscow representative of the Kazakhstan Bank and was soon making deals with all the gusto of the new biznesmeni class. He liked his new life but noticed the local militia had tapped his phone and was logging his whereabouts from a newly constructed kiosk right beside his apartment.

In June Alibekov was offered citizenship and a job with the Kazakhstan Ministry of Defense. He thought it might be to coordinate Kazakhstan's new army medical unit, but there was a hitch. He was offered the job of overseeing Stepnagorsk, in the career he had grown to despise. He knew General Kalinin was behind the offer and that the joint Russian Kazakhstan agreement to continue making weapons was in violation of the Biological Warfare Treaty.

Colonel Alibekov was furious, but he also knew he was trapped. Unless he accepted the role picked out for him by people like General Kalinin, he would have no career in either medicine or research. It was his old friend Savva Yermoshin who finally showed Alibekov the way. He bumped into the KGB director at a meeting of the Russian Biological Society. Yermoshin punched Alibekov on the shoulder and asked him affectionately if he was a millionaire yet.

"I'll let you know when I become one!"

"You know Kan, some people are nervous about you."

Savva explained that he kept telling people that there was nothing to worry about. True, Alibekov traveled a lot, but he would never live in another country without his family - and of course he would never get permission to leave with his family. [9]

From then on Alibekov was clear about what he had to do. He gave Lisa Bronson's business card to a Russian businesswoman and asked her to call

Lisa from her home in New York. A few days later Alibekov called his contact while attending a conference on the island of Malta.

She reported that she had spoken with his friends and that they would be very glad to have him visit the United States. [10]

From then on, things proceeded quickly. Alibekov flew to New York to receive precise instructions for getting his family out of Russia. He returned to Russia to pick up his family and proceeded to fly to Kazakhstan to say goodbye to his parents. The family then flew back to their apartment in Moscow and had a friend drive them to an airport in another part of the city where they caught a flight to a still undisclosed third nation before landing in New York's Kennedy Airport. Alibekov could finally breathe easily. He had outwitted the KGB.

Alibekov spent his first year in America learning English, being debriefed and Americanizing his name to Ken Alibek. It was a relief to be able to talk about his career after all those years of silence. But Dr. Alibek gradually realized that his debriefers were treating his testimony as a closed chapter in Cold War history. Boris Yeltsin had issued a decree banning biological weapons research just before Dr. Alibek had defected. As far as his debriefers were concerned that was it, biological warfare was over in the former Soviet Union.

Dr Alibek tried to convince the American intelligence community that the threat was still real. He explained that the same people were still running the show in Biopreparat and the Russian Army. But the State department didn't want to hear about it. They were too busy building bridges with the new regime in Moscow.

However, things started to change when another team of experts visited Russia in 1993. This time they toured two of Biopreparat's facilities under direct military command. The results were grave. David Kelly reported that "Pokrov was the most sinister facility he had ever seen." [11] The team had seen production lines with fermenters that could produce more than ten tons of viruses and hardened bunkers that contained hundreds of thousands of eggs. The Russians claimed the eggs were there to grow viruses to produce influenza and cattle fever vaccines, but the quantities were just too large. The experts argued that the bunkers were there to produce massive quantities of smallpox, enough to sustain a strategic biological weapons system. And what if the smallpox were ever sold to a rogue state like Iraq?

The situation in Berdsk looked no better. The Russians admitted that their entire production line could be switched from producing one strain of bacteria to another in only two weeks time. That meant they could go from producing the innocuous bio-insecticide *Bacillus thuringiensis* to 256,000 liters of anthrax in two weeks flat, plus any one of their ten fermenters could produce enough bacterial agents to equal Iraq's total supply. [12]

Dr. Alibek had also been right about personnel. On April 7 1994 Boris Yeltsin fired the head of the Russian committee investigating biological weapons. The reason? General Anatoliy Kuntsevich had been caught trying to sell five tons of nerve gas to Iraq through Syrian middlemen.[13] What was worse, Yeltsin replaced Kuntsevich with General Valentin Yerstigneev, the hardliner who had ordered Alibekov's team to find evidence of American biological weapons. He joined General Kalinin, still head of Biopreparat, Lev Sandakchiev, still head of Vektor, and General Urakov, still in charge of Obolensk. The former Soviet Union had not given up biological weapons; they had simply buried them deeper in Russia's new labyrinthine bureaucracy, perhaps even beyond the sight of Boris Yeltsin and eventually Vladimir Putin. The intelligence community started to realize the seriousness of the threat of biological warfare. The genie was half out of the bottle, and the bugs and gas guys had gained a knowledgeable and enthusiastic new convert in Ken Alibek.

CHAPTER 19
The MCZ - Cambridge, Massachusetts
1994

The Museum of Comparative Zoology, fondly known as the MCZ, up until some committee or other changed its name to the bloodless Harvard Museum of Natural History, is one of the most venerable university museums in the world.

For years you would enter it through a basement side door and be instantly assailed with the heady aroma of formaldehyde. To this day the smell of formaldehyde brings back fond memories of my own intellectual awakenings in the venerable institution.

You would proceed upstairs under the stern visage of Louis Agassiz the Swiss scientist who came up with the theory of Ice Ages after tramping through his native Alps and realizing that Europe and most of the rest of Northern Hemisphere had once been under more than a mile of ice.

Harvard is famous for having at least one faculty member on the forefront of every major intellectual revolution. It is equally famous for having at least one faculty member on the wrong side of every major intellectual revolution.

The University was especially fortunate in having Louis Agassiz, who fulfilled both roles. Not only did he discover the Ice ages, but he opposed evolution and founded the Museum of Comparative Zoology to prove Darwin wrong. Of course it did quite the opposite hence the belated name change.

Despite its beginnings, the museum became a renowned repository for natural history specimens. One of its collections contained hundreds of tick pelts collected in the late 1800's from the privately owned Naushon Island and neighboring coastal areas.

One of those areas was the tip of Long Island where residents had long suffered from what local doctors referred to as "Montauk knee" a debilitating condition that presented with rashes and swollen knees.

Dr. David Pershing from the Mayo Clinic in Minnesota had a hunch that "Montauk knee" might be related to Lyme disease and knew that the MCZ and other museums contained mice pelts from the 19th Century.[1]

So, in 1994 Dr. Pershing entered the MCZ and climbed the beautiful wrought iron stairs to the Mammal Department that contained drawer upon dusty drawer of tiny little mouse pelts. And embedded in the ears of some of the pelts were deer tick nauphlii no bigger than poppy seeds.

Dr. Pershing and his colleagues snipped off a small piece of each mouse ear and to their delight, found not only dried blood that after staining revealed the presence the Lyme disease spirochete *Borellia Burgdorferi* in two specimens.

Borellia Burgdorferi is what is called a Gram- negative like bacteria. Bacteria are separated into Gram-positive and Gram-negative bacteria depending whether they stain positive or negative from a stain first developed by Dr. Hans Christian Gram in Denmark.

It says a lot about the unity of life and the ingenuity of scientists that today the way doctors test for Gram-negative bacteria is by using the processed blood of horseshoe crabs, a test first developed at the Marine Biological Laboratory just across Vineyard Sound from Naushon Island in Woods Hole, Massachusetts and tested during the 1976 Swine flu fiasco.

After completing his mouse ear biopsies, Pershing concluded that Lyme Disease was not a new disease at all. It had been in New England since at least 1894. But there had been no recorded cases of the disease because there were no deer ticks along the Massachusetts coast in the late 1800's, only mouse ticks that don't bite humans.

It was only with the return of deer and deer ticks due to reforestation that the current resurgence of the disease occurred, and it had been exacerbated by housing developments, which isolated mice from their traditional predators. Plus, global warming had expanded the range of deer ticks from the Mid-Atlantic up into Maine.

This had made deer ticks "bridge vectors" allowing Lyme disease to jump from deer and mice to humans.

It was not a particularly parsimonious explanation, and when explanations grow so complex scientists grow suspicious that a simpler theory, like that of the Ice Ages or evolution, might just tip over an existing paradigm and do a better job of explaining nature. But all that would come later.

For the meantime, ticks were biting more and more people who were suffering more and more persistent symptoms that even Pershing's new explanation of Lyme disease couldn't account for.

CHAPTER 20
The Bomblet; Washington
1999

The CIA conference room went silent as Gene Johnson passed around an aluminum replica of a Soviet-style biobomblet. It was modeled on a fragment of a bomb discovered by a US technician while he was helping Soviet scientists neutralize tons of anthrax the Soviets had buried on Vozrozhdeniya Island in the wake of the anthrax outbreak at Sverdlosk. The Soviets had been trying to hide the evidence that they had been systematically cheating on the Biological Weapons treaty since 1972.[1]

Ken Alibek recognized the bomblet as similar to the ones the United States and the Soviet Union had built in the 1950s. Both bomblets housed 450 grams of agent, exactly two pounds' worth of germs. The similarities made him think that the KGB had stolen the design from some Fort Detrick blueprints. The only difference between the two weapons was that the United States pumped Freon gas around their germs to cool them during detonation while the Soviet Union protected their germs with plastic pellets.[2]

Tom Monath was chairing this meeting of the CIA's Advisory Panel on Non-Proliferation. It included microbiologists, weapons experts and defense analysts; the best and brightest from the world of the bugs and gas guys. Tom could remember the first time he met Gene Johnson, under a downpour of dying flies in Guayaquil Ecuador. Now his hero, another one of the original disease cowboys, was on a decidedly different mission. Johnson was in charge of "Clear Vision" a top secret CIA project to make and test a biobomblet.

Dr. Johnson explained to the group that the bomblet was probably the same type of weapon that Iraq had stockpiled during the Gulf War. Both Iran and Iraq had tried to hire bio weapons engineers from the former Soviet Union, and some of the bomblets had made their way onto the open market from Russia's lightly guarded cache of biological weapons.

Johnson explained that CIA operatives had learned that low-tech terrorists were also interested in making and purchasing biological weapons. One of the terrorists was Osama bin Laden a mysterious Saudi expatriate who was reported to be training terrorists in Afghanistan.[3]

But Josh Lederberg was highly uncomfortable with the bomblet. He argued that the CIA was edging very close to violating the Biological Weapons Treaty. Johnson countered that this was the most likely weapon that Iraq would use and the CIA just wanted to know how to defend against it. Besides, the bomblet lacked a fuse and they were only going to load it with *Bacillus thuringiensis* and *Bacillus globigii* and test them at Batelle's wind tunnel outside Columbus, Ohio. They were not going to use real anthrax. Still, it came pretty close to violating both the spirit and the letter of the law.[4]

Other experts argued that, if it ever became public that the CIA had made a bomblet, it would look like the United States was getting back into making biological weapons. That would surely undermine the State Department's efforts to strengthen the treaty's accords.

But Clear Vision was not the only such project the United States was pursuing prior to 9/11. The Pentagon had changed its focus from concentrating on what nations like Iraq could do to concentrating on what "third-rate low tech" terrorists like Osama bin Laden could do. The results were not encouraging.

In Project Bacchus, the Pentagon gave scientists at the Dugway Proving Ground $1.6 million dollars to see if they could purchase off-the-shelf equipment to build a small lab in Nevada to manufacture anthrax. Officials from the Defense Threat Reduction Agency were particularly interested to see if the purchases would create a signature to alert local, state and federal officials.[5]

In less than two years the scientists bought pipes and filters from a local Lowe's hardware store, a machine for milling grain from a Midwestern wholesaler, and a fifty-liter fermenter from a European company selling equipment on the open market. By the summer of 2000 they had been able to produce two pounds of "weapons grade" *Bacillus thuringiensis* and *Bacillus globigii*, just enough to load into a Soviet style biobomblet. Throughout the operation their activities never raised suspicions or provided a telltale signature of purchases that would lead federal, state or local officials to suspect their true intent.[6]

In 2001 the Defense Threat Reduction Agency used the same lab to train commandos how to identify and neutralize such a facility without spreading any dangerous germs. The project, code named Divine Junker, had the added benefit that it would destroy any evidence that the United States had strayed so close to violating the provisions of the Biological Weapons Treaty.[7]

Other projects like "Bite Size," Back Star," and "Druid Tempest" tested the Pentagon's ability to destroy biological weapons in hardened underground facilities with the their new line of bunker buster bombs. Plans were even drawn up to use nuclear tipped missiles to penetrate mountains and destroy caches of chemical and biological weapons. Later it was suggested that the United States might do this preemptively to destroy weapons of mass destruction.[8]

Project Jefferson was even more disturbing. The Defense Intelligence Agency hired Batelle, the same company that had tested the dispersal rates of the biobomblet, to recreate the anthrax chimera that Russian scientists had produced in 1995. Again, government lawyers argued that this was merely defensive research allowable under the provisions of the Biological Weapons Treaty. The agency just wanted to know if the Pentagon's vaccine would be effective against this new kind of enhanced anthrax.[9] Ken Alibek understood the rationale, but he argued that if the United States wanted to pursue such research it should be published in the open literature so that the United States could not be accused of cheating on the treaty.

"It can't be classified. If done secretly, you can imagine the problems." [10]

Alibek knew of what he spoke, he had seen the results of the Soviet Union's clandestine research. But other people were delighted that the United States was beginning to take the threat of biological warfare more seriously. They felt that President Nixon had been wrong when he stopped research on biological warfare in 1969. They had suffered during the severe cutbacks at Fort Detrick after the collapse of the Soviet Union and the end of the war in Iraq.

One of these people was Steven Hatfill who had been fired from Fort Detrick and lost his security clearance from the Defense Department in August 2001. Another man to watch was Mohamed Atta, who had just finished flight training in Florida for a mission to occur on September 11.

At the end of the seminar Hatfill said he was "rather concerned" that neither he nor any other bio-defense expert had been invited to speak. He added, "As was evidenced in downtown Washington D.C. a few hours later, this topic is vital to the security of the United States. I am tremendously interested in becoming more involved in this area."

Could Steven Hatfill, newly hired at Fort Detrick, have sent a vial of wet vegetative Bacillus to highlight the problem so he could become more involved?

In 1999, another series of letters containing dry fake anthrax powder was sent to media and government targets including the Washington Post, NBC's Atlanta office, and the Old Executive Office Building, the same building where Steven Hatfill had presented a briefing 3 months before. The letters were written in capital letters and said:

"Warning: This building and everything in it has been exposed to anthrax. Call 911 now and secure the building. Otherwise the germ will spread." [11]

The letters contained two teaspoonfuls of fake anthrax powder. Had Steven Hatfill learned about aerosolized anthrax during his two-year stint at Fort Detrick? In his resume, updated in 1999, he added that he had "a working knowledge of wet and dry biological warfare agents."[12]

When Hatfill was seen taking several biosafety cabinets from Fort Detrick in August 2000, could these cabinets have been used to teach Special Forces commandos how to identify and secure scientific papers when they destroyed the Defense Threat Reduction Agency's anthrax lab at the Dugway Proving Ground also in August 2001? [13]

By August 2001 Hatfill had lost his security clearance, his job at Science Applications International Corporation, and the possibility of a contract from the CIA. His actions after that became murky, but sources told the FBI that he had become angry and depressed about losing his security clearance. He did fly to Florida, where he kept a storage locker and to London perhaps during the time when another anthrax hoax letter was sent to Senator Daschle in November 2001. But it is clear that the September l8 and October 9 anthrax mailings had similarities to the hoax letters sent in 1997 and 1999, and perhaps the hoax letter sent from London in 2001.

First, the two sets of anthrax letters contained increasingly refined and powerful germs. This could mean that the perpetrator hurried to grow and then send off anthrax after September 11 then grew and milled more refined anthrax to mail off on October 9th.

Second, the letters urged people to take antibiotics. Hatfill was once reported to have given Cipro to people at an isolated residence believed to be a government safe house. The safe house is believed by some FBI informants to be the location where the anthrax was made and milled.[14] The fact that the perpetrator warned people to take antibiotics indicates that he did not want or expect many people to be killed.

Was Hatfill a single rogue element, the "loner with a scientific background" that the FBI profiled early in its investigation?

When Hatfill joined Fort Detrick in 1997, it was an organization that had been described as an institution with "little or no organization, little or no accountability" and such lax security that 62 samples of biological agents had disappeared over the years including Ebola, hantavirus, SIV (the virus that causes an AIDS like disease in monkeys), and anthrax. Richard Crossland, a researcher who worked with botulism testified that after he was laid off he spent three days walking out of the lab with armloads of boxes that nobody had bothered to check.[15]

The organization was said to foster a "cowboy culture" that encouraged germ warriors to produce and tinker with anthrax in their off hours beyond the scrutiny of security and lab officials. Added to this, there was a toxic undercurrent of racism within the organization.

In 1991, on a Saturday just before Easter, Dr. Ayaad Assaad, an Egyptian born research scientist who had lived in the United States for 25 years and married an American woman from Nebraska, found an eight-page poem, in the lab denigrating his ethnicity. It drew attention to a model of a camel with enlarged anatomically correct sexual appendages. One stanza of the rambling poem read,"In (Assaad's) honor we created this beast;" It represents life lower than yeast."

The camel, it noted, would be given each week "to those who did least."[16]

The poem was signed by the Camel Club, a group of 6 unidentified co-workers who continued to make Dr Assaad's life miserable until he was laid off in 1997, the same year that Steven Hatfill started working at Fort Detrick. During their years at Fort Detrick, Dr Assaad, and Dr. Kay Mereish, a Jordanian-born scientist brought several claims of ethnic harassment to their division chief David Franz. According to Dr. Assaad, "Franz kicked me out of his office and slammed the door in my face, because he didn't want to talk about it." [17]

Why was such lack of accountability condoned and tolerated within the organization? Did administrators want more money put into the budget to counter the threat of bioterrorism?

Perhaps the last word about the anthrax mailings belongs to David Franz, the division chief who allegedly failed to stop the racial baiting of Dr. Assaad. He always said the hardest part of his job was having to lay off people during the budget cuts that followed the Gulf War. On April 4, 2002, he told ABC News, "I think a lot of good has come of it. From a biological or a medical standpoint, we now have five people who have died, but we've put about $6 billion dollars in our budget into defending against bioterrorism." [18]

CHAPTER 21
Eckard Wimmer; The Polio Maker
July 11, 2002

With his old world manners, tweedy jackets, and German accent, Dr. Eckard Wimmer looks like he should be studying arachnids in a musty lab tucked away in some obscure old corner of Heidelberg University. Instead he and his wife Astrid are a model academic couple on the Stony Brook campus of the State University of New York. He is the former chairman of the Microbiology department and she is a professor of comparative literature.[1]

At 3:00 p.m. every afternoon, Eckard and Astrid serve high tea in pottery thrown on Astrid's campus kiln. On weekends they putter about their garden and sing in the university choir. Could this be the man whose bemused physiognomy is plastered all over the pages of the New York Times? Could this be the man who is being vilified by his colleagues for being a modern day Dr. Frankenstein? Could this be the first man to create life in a test tube?

Yes, but was his creation an innocent ameba? No, his creation was polio, one of mankind's most ancient enemies that, thanks to the Pentagon and U.S. taxpayers' money, can now be used as a potentially unstoppable biological weapon.

Who is Dr Wimmer and what exactly has he done? Eckard Wimmer grew up in Nazi Germany during World War II. His father was a brilliant chemist who was killed in the German army when Eckard was only three years old. He left Germany in 1946 to study biochemistry at the University of British Columbia, then virology at the Saint Louis School of Medicine. It was at Saint Louis that Dr. Wimmer first became fascinated with polio and appalled by its unique ability to kill and paralyze its victims.[2]

In 1974, Dr. Wimmer moved to Stony Brook where he continued to study polio, brain tumors, and male infertility. In 1999, Dr. Wimmer was approached by the Pentagon. They wanted to know if it was possible for someone to create polio from scratch using a genomic blueprint available on the internet and fragments of DNA purchased from a commercial supply house.

Dr. Wimmer thought it was indeed possible and applied to the Pentagon's Defense Research Projects Agency for funding. They supplied Dr. Wimmer and two assistants with $300,000 and asked them to produce life out of nonliving chemicals. This would be a first for biology. But they didn't want him to create just any kind of life. They wanted him to create his old favorite, polio. They reasoned that if Dr. Wimmer could create a biological agent from scratch, then bioterrorists could do the same thing.[3]

As it turned out, the project was relatively easy. Dr Wimmer's team bought stretches of DNA oligonucleotides for 40 cents each from Integrated DNA Technologies in Iowa City. They then added an enzyme to convert the DNA to RNA and stitched them together. The string of RNA was then stirred into a mixture of proteins and voila, poliovirus made from scratch.

The entire process took three years to accomplish, but Dr. Wimmer thinks it could now be done in less than six months for a fraction of the cost. After the virus was created the Journal of Science published the entire recipe on the internet for the whole world to appreciate.[4]

Perhaps the most troublesome aspect of this research is not that terrorists are likely to set up shop and churn out synthetic versions of polio, anthrax, or Ebola. It would be far easier to harvest virus from a child infected with polio, or to dig up the corpse of a cow who died from anthrax, or manipulate monkey pox to create smallpox.

You could even catch a few mosquitoes infected with West Nile Virus and FedEx them to a compatriot in some congenial place like Louisiana or Washington D.C.- birds and mosquitoes would do the rest. No, the real danger lay in how easy and inexpensive it was to create pathogens and now thanks to the Pentagon, anyone with a smattering of biological training could create their own superbugs.

Some biotech companies even deliver bits of DNA in special refrigerated vending machines so researchers don't have to have to order them individually. All a busy researcher has to do is go down to his university's supply room, punch a few buttons, swipe his credit card, and voila! The vending machine will drop out a $50 vial of just the right bit of DNA he needs to finish his life form.

A sophisticated suicide bomber can now infect Jerusalem City, rather than strap a bomb to his chest; an eco-terrorist can poison an entire logging camp, rather than put nails into a few trees; a high school nerd can now give flu to the homecoming football team, rather than gun them down in the school cafeteria. It seems fair to ask if taxpayers' money should be spent on providing this sort of information to whomever might want to harm us.

But perhaps the most disturbing aspect of this tawdry stunt is that it shows how much we have perverted the discipline of biology. Biologists used to be given modest grants to unravel the secrets of the universe, to save human lives, and to provide insight into and, dare I say it, wisdom on the importance of saving both our species and our planet from extinction.

In previous decades most of the money for biological research came through two governmental agencies, the National Science Foundation and the National Institutes of Health. Today more and more money for biological research comes from the CIA, the Defense intelligence Agency, the Pentagon, and pharmaceutical and agricultural conglomerates. By 2005 the Defense Advanced Research Projects Agency is projected to have spent $1.2 billion dollars on biomedical research.[5]

While all of this research is supposed to be carried out in order to protect against biological warfare, most of it involves using gene-splicing technology to control nature, create antibiotic resistant germs, and developing new ways for killing human beings. We have created a system of government funding of civilian research that almost mirrors the former Soviet Union's Biopreparat program.

We have created a process that makes it increasingly possible that somewhere, somehow, someone is going to set up a small lab and unleash such a rapidly spreading pandemic that it brings down civilization and alters forever the course of history — if not leading to the total extinction of our species. If so, it will not be because of some evil-minded James Bond character bent on conquering the world, but probably because of some mild-mannered scientist with the best of intentions - like Albert Einstein or Eckard Wimmer.

During the Second World War, physics went through the same kind of loss of innocence that biology is facing today. In his memoirs, Albert Einstein wrote that his one big mistake was writing a letter to President Roosevelt urging him to fund a project to develop the atomic bomb. At the time, Einstein felt it was important that the United States develop the bomb to end the war before Germany did the same. But suddenly money and prestige had flowed into the discipline of physics, which had formerly been a quiet academic backwater where nerdy scientists argued among themselves about the nature of the universe.

However, as the Manhattan Project progressed it became less about ending the war and more about controlling the aftermath. It became less about saving allied lives and more about tilting the balance of power toward the West and away from the Soviet Union. The world of politics had discovered that physics held the power to alter and control the future.

Today biology is in the same position. Government and industry have discovered that gene splicing holds the power to alter and shape the future of life and our planet in ways that were unimaginable a decade ago. Already almost all the food we eat has been bathed in antibiotics or altered by gene splicing. Pharmaceutical companies have devised new ways of altering our genetic makeup, and significant amounts of money are being spent on finding new ways to control and kill humans and other species.

Ninety-nine percent of these changes may prove to be beneficial. They may allow us to use fewer pesticides and fertilizers to grow more food. They may allow us to live longer and better lives. But they may also allow us to unleash a rapidly expanding epidemic that changes the world. Ecologists like to discuss events like asteroid impacts that extinguish one group of animals like the dinosaurs to make way for another group of animals like mammals. Informally, they refer to these events as "wiping the slate clean" to make room for a new species. Do we want to be the generation that has "wiped the slate clean" to prepare the world for a species less foolhardy than our own? [6]

PART II

Covid 19

Insights and Impressions

CHAPTER 22
The Chicago of China; Wuhan
2020

To fully understand the origins of Covid-19 it helps to have an understanding of the city of Wuhan and its role in Chinese history.

The Chinese have always revered nature and been endowed with abundant natural resources. Whether it is peasants toiling in their rice paddies or mandarins puttering in gardens decorated with fishponds, stone bridges and meditation pavilions, China is a nation of nature worshipers.

The Chinese know how to gather wild food and eat fresh game, but they also know how to turn their treasure trove of natural resources into highly refined products like silk, tea, porcelain, cars and computers.[1]

It was their early valuable commodities that attracted western traders to the celestial kingdom, but there was a catch. The West really didn't have anything the Chinese wanted to buy.

Westerners tried to make up for this deficit by driving several species of seals, otters and sandalwood trees to the brink of extinction, but the Chinese remained contemptuous of their barbarian visitors and restricted them to living and trading only in foreign compounds in Canton.

Eventually Great Britain came up with a two-tiered trade strategy; first hook the Chinese on opium shipped from their Indian colony and second, steal Chinese tea and learn how to grow it in India. The two-tier system would guarantee the British would continue to have enough revenue to administer and expand their control of the Indian sub-continent.

At its peak the British revenue from opium came to $1.3 billion in today's dollars and one out of three Chinese was hopelessly addicted.[2]

The exalted emperor of the Middle kingdom finally banned the sale of opium, held all the foreign traders hostage and dumped 3 million pounds of British opium into the Pearl River. This initiated the Opium Wars, which ended when Western ships steamed up the Pearl River laying waste to fleets, junks and forts.

The Tianjin treaty that ended the second Opium War forcefully opened up China to the trade it didn't want, and one of the first treaty ports the Westerns settled was Wuhan on the Yangtze River. By 1850 Wuhan was already a thriving city boasting a million inhabitants, which was half the size of the world largest city, London.[3]

Wuhan was also China's largest inland entreport supplying coastal cities with tea, meat and tobacco. While it became a cosmopolitan city with elegant consulates and foreign trading houses, it never developed the Bohemian nightlife, art galleries and publishing houses of Shanghai or the scholarly attributes of Beijing.

But it became the center of revolutionary change. In 1811 a careless dissident soldier accidentally dropped his lighted cigarette onto a pile of incendiaries, which exploded, causing a neighbor to call the cops.

The authorities seized documents outlining plans to overthrow the Qing government. Faced with certain execution soldiers in the Wuhan garrison fought back and eventually the 267-year-old Qing dynasty came to a catastrophic end.

After the change in government, Wuhan developed into the center of China's heavy industries, producing iron and steel for the railways which were starting to spider their way through China, alongside Wuhan's traditional cotton mills, silk filatures and forges for manufacturing horseshoes which were still valuable commodities.

It was these strengths that made Wuhan a target. In 1937 Japan invaded China, bombing and raping the eastern ports of Shanghai and Nanjing forcing China's west leaning Chiang Kai-Shek to retreat up the Yangtze River and establish his capital in Wuhan, but it fell to Japan in October 1938.

Though it lost its political status, Wuhan quickly rebuilt itself to become the center of China's automobile industry financed by the likes of Honda, Citroen, and General Motors.

At the same time it became the center of China's pharmaceutical industry, and home to the Chinese Institute of Virology, housing China's only Level 4 dual use Biocontainment facility able to produce both vaccines and potentially dangerous pathogens like Covid-19.

CHAPTER 23
The SARS Epidemic
2002-2004

On November 16, 2002 a farmer in China's Guangdong Province contracted pneumonia.

The Chinese government discouraged the press from reporting on what became known as SARS, Severe Acute Respiratory Syndrome, and did not provide information to residents outside of the Guangdong province or in nearby Hong Kong.

China only got around to reporting the outbreak to the World Health Organization in February and then prevented a WHO team from entering the province for several weeks. It was only after months of international criticism that the Communist Party relaxed their press restrictions.[1]

SARS spread quickly to Hong Kong, Vietnam and the rest of Asia; an elderly woman contracted the disease in Hong Kong's Metropole Hotel and then flew to Canada where she infected her son who spread it to patients in Toronto's Scarborough Grace Hospital. SARS continued to hop scotch all over the world, eventually killing 774 people and infecting 8,000. The numbers were piddling. But for a few small biological factors it could have been a pandemic not merely an outbreak. Humankind had lucked out, as it wouldn't with Covid-19. The world had missed its huge wakeup call.

Tracing the origins of the SARS was not easy. It took researchers several decades to discover that the disease had originated in caves in Yunnan Province. It had emerged from wild bats in the caves and the researchers realized there were hundreds of similar pathogens lurking in other wild animal reservoirs that were poised to mutate into highly infectious human diseases.

World health leaders suspected that SARS represented what is called a spillover event in which a pathogen can reside in a reservoir species for decades if not millennia until it spills over into an amplifying organism. So a virus might reside in birds that live close to pigs until the virus spills over into the pigs where it mutates to increase its infectivity so it can jump from human to human. This secondary animal is called an amplifier. So when we call a disease swine flu or Eastern Equine Encephalitis we are naming it after its amplifier organism.

Such a spillover event is different from an illness like Lyme disease that needs a vector in this case a tick, in the case of malaria a mosquito, to infect a human. Because such non-vector diseases pass directly from human to human they can quickly become true pandemics.

So, in 2014, the United States, China and several other nations invested millions of dollars to train researchers how to track down such pathogens and figure out how they might mutate and lead to future pandemics. And in 2019 the U. S. Department of Health gave the Wuhan Center for Virology an additional $7.4 million in funding to see if they could make bat coronaviruses mutate into human pathogens.[2]

They did this through a process called gain of function research. The procedure involves removing a virus from a bat and injecting it into a surrogate amplifiers in this case a series of ferrets until the viruses eventually mutated so they could infect another ferrets on their own.

Ferrets make particularly good research animals for this work because they are susceptible to the same diseases as humans. So as soon as you have a virus that can jump from one ferret to another, you know that you have a pathogen that can do the same thing in humans.

You have also created the next potential pandemic. All that has to happen is for one of the sick ferrets to be sold to the nearby Wuhan wet seafood market, as was reportedly often done, or for a researcher to prick his finger or get bat feces on his hands to get infected with the new virus and be sent to a hospital for evaluation and treatment. There, he or she could infect doctors or other patients as happened in the Scarborough Grace Hospital in Canada or the Union Hospital across the street from The Center for Virology in Wuhan.

Researchers in the field are so concerned about the threat of such a common laboratory mistake causing a pandemic that two hundred of them formed the Cambridge Working Group in Opposition that in 2014 called for a ban on such research. But it was ignored and China's only level 4 Bio containment facility went online in the Wuhan Center for Virology in 2018.[3]

On January 19, 2018 officials in the U.S. Embassy reported that there was a serious shortage of trained technicians in the lab and that safety procedures were lax.[4]

On April 24 2020 the National Institute of Health cancelled Project Predict's gain of function research. It was being conducted in the Center for Virology by an American research organization called Eco-Health Alliance.

Who had funded and championed the project? Dr. Anthony Fauci, of the National Institute for Allergy and Infectious Diseases. When asked about the origins of Covid-19 he said, "When this is all over we will have time to figure out how this all came about." Did he feel a sense of responsibility for unleashing the Covid-19 pandemic on the world? He would certainly do his best to slow it down.

But the saving grace of SARS was that patients had symptoms before they became infectious. Unlike having the flu where someone could ride a bus or fly in a plane before getting seriously ill. As the world would soon discover, with Covid-19 you could become infectious before you became symptomatic, in fact you could become virulently infectious while still being asymptomatic. The world had been lucky with SARS, would it run out of luck with Covid-19?

CHAPTER 24
The Tired Technician; A Faustian Tale Wuhan, China
2019

When the Covid-19 story first broke, the right wing swung into action in a very knee jerk blame it on the Chinese kind of way. President Trump followed suit by banning all flights from China into the United States.

The left made this into the right wing conspiracy narrative and developed their own counter narrative that Covid-19 was just like any other zoonosis disease that had jumped from animals to humans.

But just because the conspiracy narrative was being touted by the right wing didn't mean it was necessarily wrong. With that in mind lets proceed:

The official story of Covid-19 is that for millennia before there were humans, coronaviruses were simply diseases of horseshoe bats. Then one day, at one time, a coronavirus jumped from a bat to a snake, or a kind of scaly anteater called a pangolin. There it mutated the ability to be transmitted from human to human. This begs the question of why such a mutation would benefit the virus when it was still in the pangolin.

And of course all of these mutations were supposed to have happened in the Wuhan wet seafood market where stacks of rabbit cages are stacked on top of civet cages, which rest on cages of ferret badgers. A vertical petri dish if you will.

The problem is, researchers looked, but never found Covid-19 in snakes or pangolins and never found it in any of the other animals in the market. In other words they never found the amplifier organism. They also never found patient zero, the first person to come down with the novel coronavirus disease, who mysteriously disappeared.

Their alternative scenario was that the virus first jumped from animals to humans and then later developed the ability to be transmitted from human to human, in a human. This means there would be less chance of the virus evolving once again. If it had only made the jump once it meant that it could happen repeatedly and we would have recurring outbreaks. In this case humans themselves would be the amplifying organism. This is all getting pretty complicated, so lets back up a bit.

Viruses come in two flavors, DNA and RNA. DNA flavored viruses include Herpes, shingles and Papilloma viruses that all remain in your body for life. But RNA viruses are simpler and mutate more rapidly and can retain their advantageous traits. But they only remain in your body while you are sick and they then have to find another host or they will die out. So RNA viruses tend to cause rapidly spreading epidemics like Ebola, Zika and the flu.

Coronaviruses are RNA viruses that are covered with spikes that give them their name. They include the recently discovered SARS and MERS viruses as well as the viruses of the common cold.

But the difference between SARS and MERS and the cold viruses is that they have evolved specialized hook and cleavage structures. These structures allow the viruses to lock onto a cell membrane, then tear it apart so the viruses can insert their RNA into the cell. There it highjacks the cell's genetic machinery and the cell will start churning out copies of the invading virus to spread to other hosts.

The other crucial difference is that cold viruses only affect the upper respiratory tract where they make you feel miserable but don't kill you, but the SARS and MERS viruses sweep down into your lungs where they can, and often do, kill you.

So Covid-19 has both the virulence of SARS and MERS, and the communicability of the common cold. Now, it is possible that nature could merge the characteristics of all these different coronaviruses in several different species on its own, but it is so unlikely that a scientist might call it a "just so" story. So if nature didn't create Covid-19 who did? Lets look at a third scenario.

The Tired Technician

Suppose there was a diligent technician working on the other side of the Wuhan dual use facility, the lab where they make and stockpile samples of biological agents. The agents would be stockpiled there so that in the event of a biological attack, researchers could use them to create a vaccine. Of course the agents could be used to make illegal biological weapons as well.

Now suppose that the technician requisitioned some bats from the Center of Virology's bat colony that housed over 600 bats. Then all he, or she, would have to do is use the new CRISPR gene splicing techniques to snip a gene for communicability out of a common cold virus, snip a gene for a hook out of a SARS virus, snip a cleavage gene from a MERS virus and voila he would have a virus which was both communicable and able to work its way into the human lung. Or suppose he was doing gain of function research and was injecting a series of ferrets.

It would take the technician several months to perfect such a pathogen and he would have undoubtedly been exhausted when he was finally done. Perhaps he was careless while he was putting one of the ferrets back into its cage and it bit him on the hand. Perhaps the ferret wiggled and the technician slipped and jabbed a needle into his hand. There are records of similar incidents happening in the Wuhan Center.

Now suppose the technician stanched the blood and reported to his superior who sent him to the adjacent Union Hospital, which was where the first group of doctors was infected. But the doctors were unaware of the new disease so they passed it on to their patients who were also among the first people exposed.

It is also possible that the head of the Wuhan Center for Disease Control, Wang Yanji, sold the technician's bats to the neighboring Wuhan Seafood Market. One of her subordinates claimed Yanji had brought in more than a million dollars to the lab by selling research animals to the seafood market where 40 of the original patients were also found.

The patients in the Union Hospital and the Seafood market have all been accounted for. But so far, nobody has been able to track down either the amplifying organism or the first patient who started the pandemic. Could patient zero be our tired technician, and it was he, not nature, that created Covid-19?

Scripps researcher Christian Alexander argues that this would be unlikely because such a pathogen wouldn't make a very good biological weapon. But he is probably not very familiar with the origins of biological weapons.

When the U.S. Army introduced the idea of using biological weapons to President Eisenhower, they made the argument that biological weapons were the humane alternative to conventional weapons precisely because they didn't kill people. They would let you incapacitate a county like Cuba long enough so you could capture it with your conventional forces.

It was Ike who asked the pertinent question, "But happens if your bugs blow back on your own troops and spread to civilian populations." Unfortunately what Ike didn't foresee was such bugs spreading all over the world.

Isn't it just as compelling to make the argument that the technician could have actually designed a perfect biological weapon, one that would spread terror but cause relatively few deaths and be difficult to tell from a natural disease — or that the technician was doing gain of function research and made a simple mistake?

So the question remains, how likely would it be that the technician accidently infected him or herself and spread the new virus into Wuhan and the world? This would make the technician patient zero instead of an innocent shopper at the Wuhan Seafood market.

It is also likely that the technician's superiors would want to keep his existence secret, because it would reveal that they were not abiding by the 1972 and later chemical and biological weapons bans along with most other countries that have such dual use facilities.

Scientists are trained to look for the most parsimonious solution to a problem and to many of them the theory of the tired technician makes a far simpler and more credible solution than the one being touted as the official explanation.

CHAPTER 25
Fangcang
The Chinese Solution
2020

On December 30, 2019 Chinese scientists used a Polymerase Chain Reaction to determine that one of the pneumonia patients in Wuhan's Union Hospital had a coronavirus similar to the one that caused the 2002 SARS epidemic. Little did they know that their tiny little polymerase reaction would trigger a political chain reaction that would reverberate tsunami-like around the world.

Back in 2003, The Chinese Communist Party had lost face because of its slow response to the SARS epidemic. In its aftermath they vowed to be ready for the next pandemic.

So when the ophthalmologist Li Wenliang used his WeChat account to warn his friends and colleagues about the discovery of new coronavirus the post was leaked to the Wuhan police who questioned and charged him with rumor-mongering. It was still December 30th.

The following day, Wuhan officials reported the outbreak to Beijing and issued a health alert that indicated that all the 44 patients had probably picked up the virus at Wuhan's wet seafood market.

Were they like the generals who always fight the last war? No not really, their suggestion made a certain amount of sense. SARS had been traced to live wildlife markets that had sold civets that had been bitten by infected bats, so the officials brought in the usual suspects and closed down the Wuhan market. Then Beijing reported the outbreak to the World Health Organization, a mere two days after the virus had been detected. It was still January First.

Because of their recent experience with SARS Chinese officials instantly recognized the threat of the new disease.

So, twenty days after detecting the virus the Chinese Communist Party locked down Wuhan and quarantined 36 million people. They knew they had to cut off transmission and protect their hospitals from being overrun so they solved both problems with what Richard Horton * the editor of Lancet called a deceptively

---

As written in "The Covid-19 Catastrophe" by Richard Holton, Polity Press 2020.

simple innovation, widespread testing combined with a system of temporary facilities called Fangcang shelter hospitals, which sounded like Noah's Ark to Chinese ears.

People were randomly tested on Wuhan's streets and if they exhibited any Covid symptoms they would be rounded up and placed in one of the 16 hastily built Fangcang hospitals that could care for 4,000 people.

In the shelters, nurses provided their patients with food, water, intravenous fluids and basic medical service while monitoring all their vital signs for changes.

If a patient's health deteriorated doctors would transfer him to one of the regular hospitals where he could get more intensive care. But 80% of the patients only exhibited mild symptoms and were well cared for in the Fangcang facilities.

The system not only stopped the virus from being spread in workplaces and public areas but it also stopped the virus from being transmitted within families which was considered to be the most likely route of infection.

The combination of enforced testing and isolating patients in shelter hospitals may have sounded draconian to Western ears but it worked spectacularly well.

In the United States, a few universities and cities built similar shelter hospitals but the facilities remained largely empty because people were not being adequately tested and forced to quarantine.

After 9/11, one clever journalist writing about Rudolf Giuliani said that there are just some times when you want to have a paranoid control freak in charge.

Perhaps the good people of Wuhan thought the pandemic seemed like a pretty good time to have their control freak Chinese communist party in charge as well.

Chapter 26
The Importance of Being Human
Colombia, South Carolina
February 29, 2020

The day before Trump gave one of his infamous Coronavirus press conferences, the Democratic candidates held their 10th debate. Both events showed just how inextricably intermixed Covid-19 would become with race and Presidential politics.

The United States was in an existential crisis. Markets had crashed, the virus was exploding and testing and preparedness were nil. In the words of the comedian Colin Jost expecting the President to calm you down was like expecting cocaine to cure insomnia.

People become like a troop of monkeys in such a crisis. They don't want policy wonks or identity politics, they want someone they can trust and relate to. They look to body language, familiarity, humor, self-confidence, electability, even being Presidential. They are looking for the most primate-like of primates the most humanlike of humans.

It was in this milieu that Joe Biden finally found his voice. He suddenly became trustworthy, relatable, self-deprecating and funny.

During the debate he became understandably irritated when the journalists gave his rivals more time to respond. It lit a fire under his lectern.

"I'm the only one who is stopping when my time is up."

"We appreciate that and will give you time to respond during the next segment."

"Well, it all comes from my Catholic training."

Not only was the throwaway line relatable and amusing, but it also sent the subliminal message that this was a guy who knew how to follow the rules and stay above the fray.

Joe Biden suddenly assumed the bearing of the naval officer he had once been. The kind of officer you could feel comfortable talking to about your most intimate problems but also the kind of officer you could trust to weather a crisis and return you to port safe and sound. Someone who was strong, kind, bold and fair.

Black voters in South Carolina were the first people to realize this. Barack Obama had recognized these qualities in Biden and black voters were quick to remember his prescience.

Bernie Sanders might offer blacks a more just society, but Joe Biden was in the words of South Carolina Representative James Clyburn, "a good man who had endured a lot".

So with coronavirus on their minds and Clyburn's endorsement in their hearts, the voters of South Carolina gave Biden the momentum he needed to sweep into Super Tuesday.

It felt like black voters had saved our country from an increasingly erratic president and I wanted to thank them for it. But I live in a lily-white suburb in which blacks are few and far between. I thought of calling one of my black friends from college but nixed the idea of because it just seemed too liberally awkward.

But the good thing about being a journalist is that it gives you license to ask impertinent, sometimes cringe-worthy questions. So the night after Super Tuesday I found the chance to make a total fool of myself. It was in, of all places, our local YMCA. I was sitting in the steam room surrounded by clouds of billowing steam and all of a sudden a black man walked in.

We greeted each other and I asked whom he had voted for.

"That's a provocative question!"

"I know I'm sorry."

"Fair enough. Actually I didn't vote."

"Shame on you!"

"I know I was just too busy to vote. But if I had voted I probably would have voted for Bernie. I'm 32 and most of my friends were going to vote for Bernie as well."

After having a long involved talk about his work sequencing genomes for a start-up company in Cambridge I apologized again for asking such a provocative question. He laughed and said he didn't mind.

There went all my assumptions about monolithic black voters. Here was a black guy who identified more with his generation than his blackness. Here was a black guy who identified with his work as much as with his ethnicity.

In fact he was just as complicated and contradictory as anyone else. He was not just a black guy he was just another guy, and that's the whole point, isn't it?

In the strange way that these things often happen, Covid-19 had made it clear that we we're all in this crisis together. South Carolina's black voters had shown us the way to remove our increasing dangerous leader. And it would take covid-19 and another crisis before we could return the favor.

CHAPTER 27
The Reassessment
March 27, 2020

In Late March the United States reached the dubious distinction of having more Covid deaths than combat deaths in World War I, World War II and certainly Vietnam, when the average count was only 11 deaths a day.[1]

Then on March 27, the Defense Department came out with a reassessment of what they thought constituted the origins of Covid-19.

They revised their earlier assessment that Covid-19 had definitely occurred in nature; to include the possibility that the novel coronavirus had come about because of an accident at the Wuhan Center for Virology. It was an astonishing reversal apparently based on intercepts of communications between Chinese officials immediately following the outbreak of the disease in Wuhan.[2]

Such accidents are common. The Cambridge Working Group in Opposition had catalogued over 200 accidents in labs in the United Sates, China, Russia, Singapore, the United Kingdom France, Germany and the Netherlands. One happened on average every two weeks a year.

Most of the incidents occurred when a researcher made a simple error like pricking himself in the finger while injecting a wiggling bat or mouse. Other accidents occurred when equipment broke down or power was lost during storms.

One of the worst offenders was the Plum Island Animal Research Center. They had an insectary that housed thousands of potentially infected ticks. As discussed earlier, it is believed that some of the ticks escaped and spread Lyme and other tick borne diseases up and down the East Coast on migrating shorebirds.

The working group catalogued 9 accidents at the Plum Island Center when scores of sheep, steer and pigs had to be slaughtered, once during Hurricane Bob in 1991 and another time when the lab lost power after an osprey hit an overhead electric wire. This caused freezers holding diseases like anthrax, Foot and Mouth and Rift Valley virus to thaw. The stench was intolerable as the viruses oozed out of the freezers to infect animals and workers alike. Congress decommissioned the dangerous lab and built a new one in Manhattan, Kansas presumably to be closer to its bovine constituents.

The reassessment made the case for Covid-19 by going back to the SARS epidemic when it took Chinese researchers over a decade to trace the virus to bat caves in Yunnan Province. So it was not surprising that the Chinese Academy of Sciences had initially told the world they couldn't tell whether Covid-19 had occurred naturally or because of a laboratory accident.[3]

The United States should have listened more closely instead of embarking on a cultural war that pitted President Trump, who blamed the Chinese, against the mainstream media, who insisted only right-wing conspirators could ever believe the virus was due to an accident.

CHAPTER 28
Horseshoe Crabs and Covid-19
The Life You Save May be Your Own!
March 28, 2020

In mid-March the first horseshoe crabs started to emerge from Plum Island Sound in Northern Massachusetts. But this year they would have a special significance for a planet riddled with the insidious new Covid-19 coronavirus.

The first crab to be seen was a female crab. I had cataloged her the year before and named her Barnacle Babe because her shell was covered with a dozen barnacles. The barnacles had rubbed off during the winter but had left their telltale white rings.

Barnacle Babe may have been resting after coming ashore to lay her eggs during the last new moon of March. She would be doing it again during the full and new moon high tides of April, May and June.

Horseshoe crabs have done this since well before birds, fish, mammals and dinosaurs roamed the earth. The only creatures on land at the time were mosses, ferns and giant dragonflies with 3-foot long wingspans all destined to become the coal that powered the industrial revolution and filled our atmosphere with heat trapping carbon dioxide.

Watching horseshoe crabs spawn is one of my favorite rites of spring. But this year it would have particular significance because their blue copper based blood will be used to make sure that the vaccine and antibody tests for Covid-19 are not contaminated with Gram-negative bacteria, which are as lethal as they are ubiquitous in nature.[1]

This unique property makes a quart of processed horseshoe crab blood worth $30,000 and a single crab worth $3,000 if you keep it alive and just use it for its blood. If you chop it up and use it for bait it is only worth thirty cents a pound.

We have seen that the crabs' unique life saving ability was first exhibited during the 1976 Swine flu disaster when thousands of people developed crippling neurological problems after "every man, woman and child" were vaccinated against the flu that never materialized.

So, if you find a horseshoe crab on its back, flip it over so seagulls will not tear out its book gills. Use the edge of its shell, it won't bite or sting. Its tail might appear to make a far more inviting handle but using it might cause the crab to hemorrhage blue blood. You will have saved a life and it could be your own.

CHAPTER 29
Palm Sunday; Ipswich, Massachusetts
April 5, 2020

Palm Sunday started with the cries of a pair of courting osprey wheeling and diving overhead. Our species was approaching the peak of the Coronavirus Pandemic but here was nature heralding a new day.

Tidepools glittered with the golden shards of the rising sun. But last night's winds had rolled the dunes thirty feet back, burying the boardwalk under four feet of new sand.

The storm had also flattened the beach, which was gullied and strewn with piles of straw lifted off the neighboring marshes by last night's ten-foot high tides.

But the beach was now tranquil and the tidepools reflect a more cobalt sky, noticeably lacking the contrails of aircraft thundering toward the achingly empty Logan airport.

A smattering of people walked along the shore in quiet awe of nature's power to rework the land during a single storm.

It was shocking to look across the sound and see no one walking on Sandy Point. Massachusetts had closed its parks and beaches to ensure social distancing and the storm had undermined another stalwart house on nearby Plum Island.

Waves had kicked surf clams out of the offshore sediments and strewn them on the shore where seagulls were trying to figure out how to fly off with such heavy larder. I was happy to relieve them of their burden. I would add it to my stash of oysters and mussels untouched by human hands. This was my new coronavirus shopping.

Gradually I noticed the quiet whistles of a pair of piping plover, exuberant they had survived the ravages of last night's storm. They picked through the piles of wrackline and explored the new runnels gullying the beach as their ancestors had done for hundreds of thousands of years.

Suddenly it dawned on me. It is now these fragile piping plover that seem so resilient and well adapted to their environment. And it is we humans who seem so ill equipped to survive this viral storm. Are we now the more endangered species?

Perhaps if we learn how to mend our ways and live more lightly on the earth we too may survive for another ten thousand years. Is that the lesson that Covid-19 and Nature were trying to tell us on this beautiful Palm Sunday spring morning?

CHAPTER 30
The Pink Moon; Eagle Hill, Massachusetts
April 6, 2020

On April 6 I woke up with the full moon shining directly in my face. This would be the largest and brightest full moon of the year, because the moon was now orbiting 24,000 miles closer to the earth than usual.

The sun was on the opposite side of the earth and would rise at exactly the same time as the moon set. Together these two celestial bodies would cause today's ten-foot high tide and minus one inch low tide.

Even our sister planet Venus, riding just above the moon, was adding her subtle tug to the tidal bulge circulating around the earth. Yet Venus is a beautiful but tragic planet, so full of life and promise in her youth, now feverish with global warming in old age.

We humans were now gripped with our own Pandemic of sickness and fevers and I was trying to live in accordance with the earth's ancient rhythms. So I decided to take advantage of the minus tide to return some empty shells to the oyster beds so their next generation would have a suitable substrate to settle on.

The moon was setting as I tossed my trashcan of shells far out into the water. I realized if I hurried I could still catch the sun as it rose out of the Atlantic Ocean.

I arrived at Crane's Beach just in time to catch the other side of this celestial drama. The sun had already risen as a friend and I mounted the boardwalk that led to the waiting shore.

We noticed the azure color of the ocean against the distant strand. But something wasn't right. The shore was mountainous and indented with coves, bays and craggy points. It looked more like the Amalfi Coast than the gentle contours of coastal Massachusetts.

I took a few more steps and realized we had been scammed by an optical illusion. We were looking at clouds above a strip of preternaturally blue sky. My companion insisted the sky was the ocean until she too topped the boardwalk and saw the ocean's blues and greens, far more vibrant than pre-pandemic times.

We could see the results of our celestial companions' nighttime work as well. Currents had scoured a long thin tidepool encompassed behind a sandpit that paralleled the shore.

Now the incoming tide was flowing over the sandspit to create a strong shallow current. But waves were also entering the opposite side of the pool and slowly working their way up against the tide.

The currents had created a maelstrom of rapidly swirling water topped with the two-foot high waves. I realized we were seeing a tidal bore similar to those in India and the Bay of Fundy.

But my main preoccupation was the surf clams that the maelstrom was kicking out of the sand. I wanted those clams! But a seagull had the same idea. I tore off my shoes and was about to grab a nice big fat sea clam when the current carried it into deeper water to the delight of the squawking seagull.

I decided to continue up the beach and return when the tide had dropped enough so I could reach the clam, which I felt was rightfully mine.

But I didn't want to be burdened with my shoes and a few clams we had already gathered so I placed the clams on the sand and rested my shoes on top of them to dissuade the seagull who was watching my every move.

We continued down the beach strewn with the shells of sand dollars broken by last night's storm. But it was difficult to concentrate on the sand dollars when I was sure that either a rogue wave or that seagull was about to take off with dinner and my shoes

As I hustled back up the beach I admired again how the ocean sparkled with its new intensity. Last night I had shared a photograph of the Himalayan Mountains shorn of their usual mantel of hazy pollution.

The day before the Royal Observatory of Belgium reported that our planet was no longer quaking and trembling from the frenzied activity of 7 billion human beings walking, running and driving over our planet or using heavy machinery to extract minerals from her earthly body.

The Pandemic had offered our planet some momentary relief and it was heartening to catch this glimpse of just how quickly nature can heal herself if given half a chance.

CHAPTER 31
Covid-19
And the Evolution of Planetary Consciousness
April 20, 2020

In April thousands of people started to download images of improvements to their local environments. They were not acting like the centralized brain of an advanced organism but like the network of interconnected neurons of a jellyfish, coral or octopus.

This was appropriate because our species is still living on a primitive planet slowly evolving a planetary consciousness. The process happens by fits and starts and the Pandemic was providing one of those fits ... or perhaps one of those starts.

Such consciousness will decide whether we can survive as a higher, technologically advanced and enlightened species, a species able to protect our biosphere, much like an organism launches an acquired immune response to protect itself from contagion.[1]

The images revealed the possibility. In China mammoth machinery stopped clear cutting forests and strip mining coal. They had slowed work on their belt and road initiative to tie together the entire Eurasian continent in a common market for consumer goods. Some Chinese workers in Italy may have even introduced Covid-19 to Europe.

In Texas, wildcatters stopped drilling for oil and fracking for gas. A few of their companies shuttered their offices and announced they would not reopen as demand for fossil fuels had dwindled to a mere trickle.

Russia and Saudi Arabia were competing to cut their oil production to drive the United States out of the business. And it was working, refineries and shale oil operations had lain off workers. OPEC finally agreed to cut production by 9.7 million barrels a day before inventories of oil fell to critical levels.

Seven billion people had stopped working, driving, flying and consuming material goods. The results were dramatic. The usual haze of pollution had lifted above Beijing and Los Angeles. The skies had turned a beautiful blue and were no longer crisscrossed by polluting airliner contrails.

In India and Nepal people could see the towering peaks and craggy faces of the Himalayas for the first time in their lives. It was a moving sight.

The monitor on top of Hawaii's Mauna Loa Mountain revealed that atmospheric carbon dioxide had dropped for the first time since researchers started taking measurements in 1953.

In South Africa a pride of lions had started to bask in the sun on a road meandering through the shuttered Krugger National Park. Nobody had ever seen them do it before.

In Llandudno Wales Great Orme Kashimirri goats came down from the hills to walk the empty streets and graze on people's hedges and gardens, for the first time in anyone's memory.

In Chicago curators of the Shedd Aquarium felt sorry for their penguins who were getting lethargic from lack of stimulation so they let them wander through the empty aquarium. One of the fortunate penguins named Wellington wandered into the Amazon exhibit and his head spun with wonder as he looked into the tropical fish tanks whose fish looked back with equal interest.

In Rome, the Pope intoned that the Pandemic was God's way of telling humans we had to curb our lust for material things and lead more spiritual lives. He was named after the acclaimed Saint Francis who had himself communed with nature and was known for ringing the bells of his duomo to exhort his flock to step into the streets to observe the shining full moon.

Were such things the beginnings of a new global consciousness or the death knell of yet another planet that had achieved technological dominance without its concomitant wisdom? Only time would tell.

CHAPTER 32
Three Days, Three Crises.
April 13, 2020

On April 13 President Trump held the most disturbing press conference to date. East and West Coast governors had created regional pacts to coordinate re-opening their economies. Trump thundered his distress.

"The president calls the shots. I have total control."

It signaled the beginning of a potential constitutional crisis that could have made his impeachment seem trivial. He closed the briefing by badgering the press for another full hour.

Seasoned pundits called it the worst presidential meltdown they had witnessed during the past six administrations.

Two days later Trump impetuously stopped funding the World Health Organization that was the sole organization capable of coordinating a global response to the runaway health crisis. The issue was how China had handled the origins of the coronavirus.

More information had come out that the virus was probably not the result of natural selection in the Wuhan Seafood Market, and that China should have been more forthcoming about its origins.

It was clear that the U. S. government might have to deal with a health crisis in the White House as well as in old age homes.

But Trump had surrounded himself with so many "yes men" that there was nobody left who could rein him in. Dealing with Trump was like dealing with the queen of hearts whose answer to every question was, "Off with their heads!" Trump's was simply "You're fired!"

Was this the way the world would end, with the earth's most powerful country ruled by a certifiably insane leader who was becoming a danger to himself, his country and the world?

The day after Trump's press briefing, Gita Gopinath, the head of the International Monetary Fund said, "The magnitude and speed of The Great Lockdown is unlike anything we have experienced in our lifetimes."

He stressed that "The level of GDP will remain below the pre-virus trend with considerable uncertainty about the strength of the rebound. Stimulating economic activity is more challenging given the required social distancing and isolation policies."

Of course we all remembered that the best way to lift the world out of a global depression was to start a world war. So were we at a turning point? Would Covid-19 convince our species to follow our better angels and develop a global consciousness or would we follow our darker angels and descend from Pandemic, to depression, to war?

Would we wipe the slate clean so our world would revert to a meagerly populated planet in a pre-Industrial state? Would our planet rid itself of our pestilence and continue as a peaceable planet devoid of human beings?

Or would this too pass? Would the Pandemic peak then disappear? Would our democratic institutions allow us to rid ourselves of our wayward leader? Would we right our ship of state, develop a global consciousness, rebuild a non-polluting infrastructure and get through this bottleneck to become a truly sustainable planet? We would have to see.

CHAPTER 33
No, We're Not All in the Same Boat
April 20, 2020

We had been hearing a lot of the expression, "We're all in the Same Boat". And while it sounded good, was it really true?

My life hadn't changed very much during isolation. I still wrote alone everyday, seven days a week. I just wasn't getting paid for doing so.

Since I couldn't operate a cash register and lacked the training to be a first responder, I donated my scribblings to our local newspaper, a first rate rag at that!

I was fortunate to live in the town of Ipswich that provided those of us over a certain age with a free shellfish license so I hadn't had to go food shopping in almost two months. My geezer license had allowed me to get all my protein from mussels, oysters, crabs and clams while enjoying the eternal beauty of nature.

I was grateful that Ipswich and the Trustees keep the best beach in Massachusetts open to residents during the weekends. It gave my life definition. I may have lost my income but I was not going to starve.

The rest of the week the beach's sole occupants were birds, beasts and fish. They seemed to prefer it that way.

I had lost a friend and a close colleague I was looking forward to working with and was fearful every time a member of my family had to go to the hospital for regular treatments.

I had to postpone an operation to improve my vision for driving, but that didn't make much difference. I had nowhere to go.

Other people weren't so fortunate. They were the real heroes. They risked their lives in hospitals, cop cars, markets and liquor stores. They wore masks all day and had to disinfect their hands after serving every customer. They came home with chapped faces and bruised raw hands. After their exhausting days at work they had to return home and isolate to protect their children.

You could see it on the shell shocked faces of women trying to balance work and home life, and in the haunted eyes of newly unemployed men lacking the income, purpose and camaraderie of their former jobs.

You saw them walking sad and lonely along streets devoid of cars or trapped in tension filled homes. Many had watched as loved ones died alone and afraid without the comfort of someone to hold their hands as they took their last rasping breath.

So no, we were not all in the same boat. Some of us were fortunate others had lost the love of life itself. But we were all in this together and we all needed a helping hand, even if it was only a virtual one. What a strange world we were now living in.

CHAPTER 34
The Really Big Short
April 23, 2020

On April 23 I wrote, "If you want to make a bundle, all you have to do is drive down to Cushing Oklahoma. Ask for a thousand barrels of oil which had been selling for $50,000 before the Pandemic. The nice guy at the pump won't take your money. He will pay you $37,630 to take the oil off his hands instead". Of course then you would be stuck with five tanker trucks full of crude that nobody wanted to buy, because the market was expected to continue dropping to minus $100 a barrel.[1]

But if you waited a few years you would be able to make a $137,000 profit for zero investment, because the price for a barrel of oil was expected to rise to a hundred bucks a barrel. That's just how the oil market works.

So what were the market conditions that make oil so prone to such short selling schemes? At the moment it was the confluence of the Pandemic and the collapse of the world economy.

Nobody wanted to be stuck with oil because people had stopped driving, working, flying and going to bars, restaurants, concerts and sporting events. The global economy had come to a jarring halt.

The drop in oil prices had been instigated by the collapse of the world economy, but did it also foretell the demise of the fossil fuel economy?

Alexandria Ocasio-Cortez thought so. The New York Congresswoman posted the tone-deaf tweet, "You absolutely love to see it. It's the right time for a worker led mass investment in green infrastructure."[2]

She soon realized her mistake, deleted the comment and tried to expand on her ill-advised use of shorthand. The few people who had actually read her Green New Deal knew that "worker led" didn't mean socialists running through the streets but capitalists quietly using the trillions of dollars wrapped up in workers pension funds to invest in wind turbines, photo-voltaics and the digital and physical infrastructure to support them.

The process had already started. It was the market that would ultimately replace the volatility and expense of petroleum with the stability of green energy. It would follow the same path as coal.

CHAPTER 35
Another Year Without a Summer?
April 25, 2020

April 25 dawned clear and bright with cobalt blue skies and unusually cool temperatures. We had seen the atmosphere cleansed of pollution in only three short months, would we now witness the earth cooling down?

Nobody had anticipated that carbon emissions would plunge in such a dramatic fashion. Would we see our planet cool if the Lockdown continued for three more months, 6 months, a year? Would we have a year without a summer? That would drive the climate modelers crazy!

The last time the world experienced "The Year Without a Summer" was in 1816. Mount Tambora had erupted in Indonesia spewing tons of sulfur dioxide and ash into the air. The world's temperature dropped almost 2 degrees.[1]

Crops failed throughout the Northern Hemisphere, starving Napoleon's troops as they retreated from Russia and killing New England's crops with hard freezes throughout every month of the year.

"The Year Without a Summer" was the reason the maritime artist James Turner painted such lush red sunsets and the weather was so abysmal on Lake Geneva that Mary Shelley stayed inside and wrote "Frankenstein", the classic cautionary tale about interfering with nature.[2]

But we were living in a different cautionary tale. The Lockdown had created a real world experiment. Would the world cool down, warm up, or do one then the other? Were we entering into what Seventies scientists called the Mad House World?

It would seem so. Human caused carbon dioxide is the principle driver of global warming. But how much would the drop in emissions caused by the Lockdown affect the amount of carbon dioxide in the atmosphere? Unfortunately not much.

Carbon dioxide will continue to increase in the atmosphere but just at a slightly lower rate. This is because nature's seasonal additions and removals of carbon dioxide far outweigh the addition of carbon by human activities.

For instance, the 2015-2016 El Nino event was so strong it dried up tropical areas so they were only able to remove 30% of the emissions from fossil fuels instead of the 50% they sequestered on an average year.

2020 had been expected to be an average year with a large amount of CO_2 build-up. But the Lockdown had reduced the amount of emissions in China by 25% because the Chinese stopped burning coal and global emissions had declined by only 8%.

So the amount of carbon in the atmosphere had actually continued to rise in 2020. The average for April was the highest concentration in over 2 million years. The year would reach its regular maximum in May then decline as Northern Hemisphere forests returned carbon to the earth's regular minimum seasonal amount in September.

So the Lockdown would not have enough of an affect to substantially slow global warming but it would have other effects.

For instance, the reduction in aerosol pollution could cause a temporary warming as well as affect how efficiently the oceans and forests remove carbon from the atmosphere.

And the reduction in particulate air pollution, which had made the skies so blue during the shutdown, could result in cooling, as would the increase in aerosols as they had when Tambora erupted during "The Year Without a Summer." [3]

I just knew that April 25 was the only day I could enjoy Crane's Beach. We expected to have rain and snow the rest of the week.

I had been coming out here every chance I could get to scratch up surf clams for my Lockdown chowders. But the tide was too high for digging clams so I decided to explore an empty path that led to the Crane Estate.

I was accompanied by a flock of turkeys that were reclaiming the estate as their own. I walked out onto a broad strip of green grass that undulated almost a quarter of a mile toward the distant ocean. The narrow blue strip of ocean layered on top of the brilliant expanse of green looked like a perfect abstract painting.

But it was the empty hulk of the castle and the antiquities that lined her Grand Allee that made it feel like I was walking through the remains of an ancient civilization, a civilization whose inhabitants had disappeared in the wake of a mysterious contagion. I felt like I was the protagonist in Mary Shelley's less well known novel about our world after a Pandemic has wiped out all humanity except for the "Last Man".

CHAPTER 36
Mutations; Los Alamos National Laboratory
April 30, 2020

Bette Korber was pleased when her Los Alamos lab's 30-page report was finally released on April 30. It showed that there were now two strains of Covid-19.[1]

The first strain had emerged in China then spread to the West Coast. The second had appeared in Italy probably from Chinese people working on the visionary Belt and Road Initiative that followed the track of the Old Silk Trade Route.

This second strain had spread quickly from Europe to America's East Coast, becoming the dominant strain in the world by mid-March. It had out-competed the first strain by becoming more contagious.

Korber had pushed her colleagues hard, both at her Los Alamos lab, Duke, and England's University of Sheffield. They had identified 14 mutations in over 6,000 coronavirus sequences collected by the Global Initiative for Sharing based in Germany. It had been a truly international effort.

But it was the D614G mutation of the virus' contagion spikes that had finally caught their full attention. It had made the second strain of the virus more transmissible in only one short month from mid-February to mid-March. It was a classic case of rapid Darwinian evolution.

West Coast clinicians had also noticed the difference. Fewer people were dying on average in their hospitals than in East Coast hospitals, apparently because their patients had been infected by the first strain. But this raised the possibility that people might be able to contract Covid-19 twice, first from one strain then from the other.

It also meant that the virus was evolving so fast that vaccines and anti-virals, which were mostly being developed to combat the first strain, might not work against the mutated second strain.

Bette Korber wrote that this was indeed "hard news" but she saw a silver lining. "Our team was able to document this mutation and its impact on transmissibility only because of a massive global effort of clinical people and experimental groups who make new sequences of the virus in their local communities available as quickly as they possibly can."

But on her Facebook page she also warned, "We cannot afford to be blindsided as we move vaccines and antibodies into clinical testing."

To make matters even more confusing, scientists at Arizona State University used, "new generation" sequencing techniques to discover a single sequence of Covid-19 where eighty-one letters of its genome were missing. A similar deletion had happened to the SARS virus making it less deadly back in 2003.[2]

The lead author of the study, Dr. Efren Lim said this could give such a strain a Darwinian advantage because it could spread more efficiently through a population but not kill their hosts and thus themselves.

Meanwhile researchers at two Harvard affiliated hospitals were working on producing a vaccine that would use a harmless DNA virus as a Trojan Horse to deliver DNA into their patients cells. There it would make coronavirus particles that would stimulate the patient's immune system to fight off future infections.[3]

Like other vaccine makers they were focusing on the antibodies to block the virus' contagion spikes. But if the virus was mutating as fast as Bette Korber's team said it was, their vaccine could be obsolete as soon as it was released. Was this deceitful virus smarter than the best minds in medicine? It appeared to be so.

CHAPTER 37
"We're the Wild West Out Here!"
Governor Evers, Madison, Wisconsin
May 18, 2020

"Open Immediately" ~ The Tavern League of Wisconsin

By mid-May states were trying to decide how and when to reopen their economies. They had a few examples from Asian countries where the lockdowns had been quick and thorough and the reopenings had been deliberate with extensive testing. The city of Wuhan planned to test all 10 million of its inhabitants to measure how many had contracted the virus whether they had symptoms or not.

Sweden had gone to the opposite extreme allowing schools, bars and restaurants to stay open, relying on the fact that its citizens would act responsibly. The results had been mixed, while it avoided the high death tolls of Spain and Italy Sweden had far more deaths than the other Scandinavian countries with similar medical systems.

But Sweden also had hotspots like Stockholm where twice as many people had died than usual. This far surpassed cities like Boston and Chicago only approaching New York's death rates.

Yet the strategy had largely worked. Swedes stayed home as much as locked down residents in other countries and had followed their own sanitizing guidelines. These measures had kept Sweden's overall death rate about 30% above normal, which was far less than Italy or Spain and the United States but much more than Norway, Denmark, Germany and Finland.

But there were several reasons why Sweden's strategy would not be a very good model for American states to follow.

While Sweden has a similar overall low density as large parts of the United States, a whopping 52% of it's population live in single person households, while The United States has only half that amount. So the virus had raged both through Spain and Italy's multigenerational households as well those of migrants in Stockholm's hotspot areas.

Sweden's example was in marked contrast to what had happened in the United States. There gun touting citizens had protested lockdowns, vaccines, and anything else government could do to protect people from suffering, contagion and death.

On May 15 the Wisconsin Supreme Court threw out the Governor Tom Evers' stay at home order. The Tavern League of Wisconsin had won and it sent the message to its happy members.

"Open Immediately!"

Within hours thousands of unmasked revelers descended on bars and taverns; pumping fists, waving hands, pounding on the tables as if the Green Bay Packers had just won the Super Bowl or that the Pandemic had ended, which it had decidedly not!

It had been judicial activism in the extreme and it would be Wisconsinites who would pay the full price.

Massachusetts had a different problem. Its neighbor was New Hampshire. Massachusetts had been the third hardest hit state in the nation. But its numbers were finally declining while those in the heartland were rising once again. She also had some of the best doctors, hospitals research faculties and medical schools in the country that had been guiding the state's gradual return to economic normalcy.

But New Hampshire was next door and the Live Free or Die state was not part of the 6 Northeastern states that planned to act in concert. She planned to go it alone opening her restaurants, bars, malls, liquor stores and beaches before anyone else.

What would happen if Massholes drove to New Hampshire to eat, party and drink discount liquor on New Hampshire's overcrowded beaches then returned to Taxachusetts at the end of the day?

Would both states have more Covid cases than they had already experienced in March and April? Only time would tell...

CHAPTER 38
Contrails From an Ipswich Porch
May 19, 2020

On May 19 I saw the first evidence of recovery. The single high contrail of a commercial jet streaked across the azure sky. Half an hour later another followed. An hour later a jet flew in the opposite direction.

They were probably two jets leaving from New York for Europe and one returning to the United States.

Far from feeling elated I found myself feeling strangely dejected. Our three month long ordeal was finally coming to an end, yet amidst the fear and tragedy it had also been a rare and valuable idyll.

We had sat on this sunlit porch watching the greening fields and hearing the susurrus of birds returning to our northern climes. We had fed chickadees, nuthatches and woodpeckers and feasted on the colors of orioles and tanagers after so many months of gray and black.

We had seen the winged courtship of woodcocks and osprey as they ascended toward the heavens emitting exultant cries and whistles.

We had heard the sibilant mews of catbirds and the sharp calls of crows intent on stealing each other's eggs.

We had watched the arrival of egrets, herons glossy ibis and ducks, the blossoming of orchards, and the emergence of invertebrates and tadpoles only to be skewered on the bills of egrets or to be probed out of the earth by the curved scimitar of a glossy ibis.

But the sky would not have been so blue, our observations so reflective, our conversations so muted if it had not been for this three month long absence of the usual hurly burly of our hyperactive economy and the cluttered minutia of our daily rounds.

More jets will undoubtedly follow. Pollution will refill the sky. Our horizons will be lost in haze again and people will return to their labors saddened by loss but thankful to have survived our self wrought Pandemic.

Hopefully we have also learned to take the time to be kind and gentle, to help others and to tread softly on our fragile planet. We had been afforded a rare glimpse of what our world could become if we learned to curb some of our most wonton ways.

CHAPTER 39
George Floyd
May 25, 2020

*"Our soldiers received the word and advanced to quickest time,
their silence and looks of sullen hate, were more appalling than
the clamour of our innumerous foe."*

~ Mary Shelley
The Last Man, 1826

On May 25 the heart of America was pulseless. Commerce had ceased. Governmentat was at a standstill. The pandemic still reigned. The President was playing golf.

Massachusetts and Connecticut had finally eased their safety restrictions. They were the last two states to do so. Fortunately the weather was cool so people had not flocked to the newly opened beaches or attended uncontrolled parties as they had in other states.

Instead, they gathered in small, distanced groups and related stories about their recent experiences. Service stations had given free gas to medical workers. A local minister expressed her sadness at having to attend so many sparsely attended funerals and dismay that cemetery workers had apparently cut down an old tree to make room for more graves.

The coach of the Harvard polo team emailed the ICU workers at a local hospital to invite small family groups to a picnic so their kids could meet, groom and ride his quiet, understanding polo ponies.

"Being a former journalist I couldn't contain my curiosity and gently asked the nurses what the last few months had been like. They told of not being able to hold patients hands as they faced their lonely deaths, 'I wanted to be able to squeeze their hands or give them a smile, but they couldn't even see my eyes... They couldn't even see my damn eyes.'"

There were rumors that workers in one hospital had even dressed a significant family member in a HAZMAT suit and snuck her into their patient's room so he would not die alone. It was against protocol, of course, just one of those things that good nurses do to make our institutions a little more human, a little more humane.

On the other side of the country a Minneapolis cop was planting his knee on a black man's neck and slowly and impassively killing him as if grinding a rat under his boot.

Three other cops stood by and several onlookers shot video but no one intervened. Within minutes the gruesome scene was on the Internet. They were extremely difficult to watch.

But Minnesotans did watch, and over several nights they held demonstrations, which police broke up with tear gas. They arrested a black CNN journalist and shot reporters with rubber bullets from point blank range.

The demonstrations spread across the country to New York, Chicago, Philadelphia, Boston, Atlanta and on to The White House in Washington D.C.

Police departments adopted similar strategies; establishing curfews then sweeping through the crowds making arrests as the curfews went into effect.

After months of careful social distancing cops and residents were shouting into each other's faces and over 10,000 people had been thrown into pestilent jails.

The demonstrations could not have happened at a worse time. Millions of people had been stuck home, broke and unemployed. They were fed up with racism, and the president's dog whistled signals that police brutality would be ignored.

Millions more people were ready to explode under the pressure of an administration that had been endangering lives, destroying our planet, obliterating our hopes, our futures.

Besides the politics of the demonstrations there was something else going on. We had been in lockdown for several months.

In other pandemics, both real and fictitious characters had given in to risky and licentious behavior, as did Prince Prospero in Edgar Allen Poe's Masque of the Red Death, or in the words of Mary Shelley's "Last Man",

"The inhabitants strove by riot and what they imagined to be pleasure to banish thought and opiate despair. It would have been useless to oppose these impulses by barriers."

CHAPTER 40
Pandora's Box
June 5, 2020

A few days after the first George Floyd demonstrations, infectious disease doctor Taison Bell was sleeping in his home in Charlottesville, Virginia when he received an urgent page. One of his patients in the ICU Unit had taken a turn for the worse.

Taison threw on his clothes and drove toward the hospital. But suddenly his heart started to pound and his palms broke into a sweat. "In the eyes of the law, I realized I was no longer a well-trained, highly skilled doctor, just another black man speeding down the highway at 3 in the morning."

"What were the chances I would be shot by a rookie cop just as afraid of me as I was afraid of him?"

It would be just another DWB, a Driving While Black violation. A colleague had told him about the pain and humiliation he felt when he was pulled over on the way to church as his two young children all dressed in their Sunday best looked on in bewilderment and disbelief. He could still feel the knot he felt in his stomach when he tried to explain to his kids what had just happened to them.

At the same time, Taison was an infectious disease expert who was well aware of the dangers of participating in the crowded George Floyd demonstrations. He counseled people who wanted to protest to wear mask, get tested and do their best to distance themselves.

After talking the situation over with his wife, he decided not to participate in the demonstrations but start an Internet forum to share his experience of being a black doctor with his white colleagues.

Meanwhile the surge was on its way. The gods had given Pandora her box of inestimable gifts but told her she was not allowed to open it. Of course she had, and as the pestilence flew out of the box she flung it back into the sea. Hope was forever lost.

Now wave upon wave of viruses, ills and pestilence were inundating mankind. The surge was upon us.

For months people had careful isolated themselves. But now even the most hypochondriac, the most anxious the most valetudinarian were letting down their guard. They were shopping, running errands, venturing outside.

And the risk takers had had more than enough. They wanted to get back into the action, to make money. Viruses be damned, they were thirsty and wanted to have fun!

The demonstrations had released all their pent up energy and made a mockery of the slow reopening of the economy. If people could congregate on the street in thousands of communities large and small why couldn't the economy recover as well?

The week after Memorial Day the number of cases had gone up in Alaska, Arkansas, California, Kansas, Mississippi and Utah. Los Angeles was seeing 1,300 cases a day. Mayor Cecetti was urging people who had attended protests to get go into voluntary quarantine for 14 days ... as if that was going to happen.

Hawaii had seen the number of cases increase on the mainland and extended its mandatory 14 quarantine for out of state visitors - a sure way to boost tourism.

By June 10, 19 states had rising numbers of cases and Arizona had ordered its hospitals to reactivate emergency medical plans due to their having reopened their economy too soon.

While cases appeared to be falling in previous hotspots like New York and New Jersey, by mid- June cases were rising in more than half of the states in America. More than 20 million people had taken ill and over 112,000 had died. The surge was in full swing.

CHAPTER 41
Don't Tread on Mother Nature
June 29, 2020

"In the face of all this, we call ourselves lords of the creation, wielders of the elements, masters of life and death."

~ *Mary Shelley*
The Last Man, 1826

By June 29, Massachusetts and Connecticut had eased some more of their restrictions. They were the last two states to do so. Fortunately the weather had remained cool so people had not flocked to the newly opened beaches or crowded parties as they had in other states.

A few days later I had to be tested for Covid-19 before having a cataracts operation. But when I arrived the nurse told me I would have to wait because my tests had not been ordered. It was a simple mistake but they would have to contact the on-call physician in order to proceed.

It gave me the perfect excuse to explain that I working on a book about Covid-19 and that one of things I was writing about was whether the Pandemic could have come about because of a simple mistake like someone being so harried that they had not ordered my test.

The nurse clicked her pen and put it down.

"I was in the army and the thing they constantly do is train you, things like click your pen shut before putting it down."

"They know that whenever something goes wrong it is because of human error, so they train you exactly what to do so that in the midst of an emergency your brain and muscle memory take over so you do the right thing."

"We know we can make lots of mistakes in medicine that's why we constantly recheck our protocols and procedures."

I told her the same thing was true in research.

"Years ago I was a research assistant aboard an oceanographic vessel and one of my tasks was to spend my four hour watch keeping our bottom sensing device calibrated so it made a graph of the bottom. For hours it just recorded a flat line that indicated the bottom of the Atlantic Ocean.

It was four in the morning and pretty tedious work so I must have dozed off for a bit. Suddenly I realized I had lost the bottom of the Atlantic Ocean, we were over the steeply rising mid-Atlantic ridge. I frantically turned knobs and toggled switches. I was finally able to get it under control by the end of my watch, so I ate my breakfast and tumbled into my bunk.

That afternoon I checked out the graph from the night before. Some wise guy had put ZZZ's where I had lost the bottom of the Atlantic Ocean."

So yes I know how easy it might be for a sleep deprived technician to prick his finger, get sprayed with bat urine or get bitten by an infected ferret.

Earlier in the summer, the director of the Wuhan Center for Virology reported that his institute had been working on three live strains of bat coronaviruses during the outbreak of the pandemic. But he explained that the institute's so-called bat doctor; professor Shi had not paid much attention to the SARS CoV-2 virus because it was only 80% similar to Covid-19.

The announcement rekindled the debate about the origin of Covid-19. Officials in Beijing had continued to back away from their initial insistence that the virus had emerged from the Wuhan Seafood market.

That left three scenarios. One, that the pandemic was the result of biological warfare. Two, that it was the result of evolution in bats, snakes, ticks and pangolins, or three, that it came about because someone accidently contracted the disease while doing "gain of function" research.

We can dismiss the possibility that the virus was the result of biological warfare research due to a lack of evidence, one way or the other.

The same is true of the second hypothesis. Most scientists agree that while zoonotic diseases are common it gets complicated very quickly when you have to involve evolution in snakes, bats, ticks and pangolins.

That still leaves the third possibility that the virus was the result of a laboratory mistake. And doesn't an 80% similarity sound pretty darn close to a 100% similarity to a layman's ear?

We know that scientists at the Wuhan Center were doing gain of function research where they would put viruses in a succession of ferrets until the virus evolved the ability to jump from human to human, as well as from ferret to ferret.

If a researcher had been using a virus with an eighty percent similarity to Covid-19, who is to say he or she hadn't put the virus in just a few more ferrets and achieved a hundred percent similarity?

After all that was what Project Predict was all about, predicting how long it would take a virus to become transmissible in humans.

If workers achieved this hundred percent transmissibility and someone had accidently picked up the wildly contagious virus and spread it to other patients at the nearby Union Hospital, isn't it understandable that China would want to keep the information under wraps for as long as possible? Isn't it likely that a researcher who had taken the Hippocratic oath would feel remorse for starting a pandemic that had already killed over 800,000 people?

These three scenarios all have something in common. They would all come about because of humans' inclination to alter our natural environment.

It is pretty easy to see the foolhardiness of offensive biological warfare. It is morally corrupt, repugnant and dangerous.

But what about defensive biological warfare, stockpiling pathogens so you could develop vaccines in the event of an attack? Isn't that a case where the cure is arguably worse than the problem?

Shouldn't both offensive and defensive biological warfare be condemned and banned to protect our species from intentional or accidental annihilation.

And if Covid-18 came about because of a technician doing gain of function research, then shouldn't such research also be banned as a permanent danger to humanity?

Finally the idea that Covid-19 came about because of natural selection is misleading. If it did evolve in the wild it would have been by artificial selection, through deforestation, practicing agriculture too close to wilderness areas, or the practice of eating bush meat. This is a form of tweaking nature, which we also continue to do at our own peril.

So all the possible origins of this pestilence show that if humans want to survive, we have to stop tweaking nature, not only because nature is sacred but because it is also a dangerous place replete with bacteria, viruses, bats, serpents and ticks.

If we want to continue to live in harmony with nature we have to learn how to respect its dangers as well as honor its importance to our continued existence on this, our one and only planet.

CHAPTER 42
The Strange Case for Optimism
July 14, 2020

By mid-July it was clear that the pandemic was out of control in most of the United States. Florida had surpassed New York as the world's epicenter for Covid-19. If it were a country, Florida would only fall behind the United States, Brazil and India as having the fourth most cases of Covid-19 in the world. It made you want to live in a nice little third world country that knew how to handle such things.

Florida was followed closely by Texas, California and Arizona but cases were rising in all the other states, except Vermont, Maine, New Hampshire and Rhode Island, which were falling, and Wyoming and Massachusetts, which were on precarious plateaus.

It was becoming evident to even the most obtuse that going to bars, restaurants, gyms and Covid parties could be lethal, so states were trying to re-shutter their doors. They were also grappling with whether to reopen schools. However many teachers had voted with their feet, seeking other forms of employment rather than risking their lives, and the majority said they would not teach in the fall.

Lined up against all this grim news was the revelation that four of the 140 companies working on vaccines had positive results in their Phase One trials. The top contender was Moderna. Its vaccine had triggered a positive immunity response in all of the 45 trial patients who had taken varying dosages of the new vaccine. Astra-Zeneca's vaccine had also activated killer T-cells that they believed would provide immunity for years rather than months.

So the U.S. Institutes of Health had given the top contenders permission to start 30,000 person phase three trials on July 27. If Moderna's trials showed positive results it planned to produce 500 million doses of their new vaccine by the end of 2020 and up to a billion doses in 2021. In fact it had already started producing them.

Meanwhile the pandemic had made Trump the most frustrated man in America. His staff were testing positive, Fauci had higher ratings, nobody was showing up for his rallies, his niece had written a best selling book calling him the most dangerous man in the world.

But Joe Biden, who had run a lackluster campaign until black voters had put him on the map in South Carolina, had steadily grown in stature as the pandemic had progressed. It had highlighted Trump's increasing erratic behavior, his mishandling of the crisis, his disdain for civil rights, Anthony Fauci, science, and common sense in general.

But Biden still lacked the support of the progressive wing of the Democratic Party, so he had asked his staff to sit down with Governor Jay Inslee's aides who had crafted a visionary plan to build green new technology for Washington State, home of some of the most forward thinking technology firms in the world.

So on July 14 Biden announced his plan to spend $2 Trillion dollars to create a new green infrastructure to carry the United States deep into the 21st Century. The plan contained the promise of having the right scope and vision to help humanity as it struggles to squeeze through its environmental bottleneck into a potentially long-lived and fruitful future. If adopted we will find itself engaged in the same rapid pace and exhilarating changes experienced by the country under FDR. As Biden put it:

"When Trump hears environment he thinks hoax. When I hear environment I think jobs."

So while it is always risky making predictions about things as erratic a pandemic or politics, I remain pessimistic about the short term but find myself guardedly and unexpectedly optimistic about the long-term. It is clear that with the present administration in charge the pandemic will continue to get worse until states reshutter their economies until safe and effective vaccines can be produced and distributed.

But we have almost reached rock bottom. People are fed up and realize we cannot go back to a mythical past. We have to drastically change our ways to save our environment, our economy and rid ourselves of our heritage of crippling racism.

But who would have thought Joe Biden would be the one to lead us into an era of such radical change? I guess Pandemics have a way of doing such things.

CHAPTER 43
Covid-19; The Vaccine Trial
July 27, 2020

On July 27 Dawn Baker rolled up her sleeve and made history. The vivacious news anchor from WTOC in Savannah became the first woman, the first African America, hell, the first human being to get a shot of Covid vaccine in this its Phase 3 clinical trials. If the results proved positive the government would start distributing millions of doses of the new vaccine to a grateful nation.

It was not the first time Baker had volunteered for such a medical cause. In 2006, she joined the Goodness and Mercy Foundation in Ghana, then flew to Guatemala to help with the group Faith in Practice. Months later she was back with Goodness and Mercy in Nigeria. In her bio she wrote.

"It is my belief that all of us are responsible for doing our part to improve the quality of life of our fellow man."

"African Americans have a history, for good reason, of being concerned about medicine. In the past we were subject to all kinds experiments where we didn't know what was going on." She was referring, of course, to the infamous Tuskegee experiments in which African American men were infected with syphilis without their knowledge or consent.

This time Dawn was volunteering to be the first person to try the new vaccine in honor of her friend Lyndsey Gough, a fellow reporter at WTOC. Lyndsey had contracted Covid in late June then spent eleven days in the hospital and lost her appendix during the process. And Lyndsey still wasn't quite right.

"It really just broke my heart. She was such a young, strong, energetic young woman who is just so fierce about what she does and to see her afterwards, she was just so weak and to also know what could have happened to her."

"Many of my friends and even some of my relatives have said, 'You don't want to be the guinea pig. Why don't you just wait and see what happens first.'"

"I hope that maybe just seeing my face will help them change their opinions about that. It could eventually save their lives."

At least 25 other vaccines were being tested on humans in other parts of the world. And five of them were going into Phase 3 testing on the same date. This was the most advanced phase of testing before a vaccine goes to market.

Baker's trials were part of what the Trump Administration was calling Operation Warp Speed. Judging from what happened when President Ford rushed "to inoculate every man woman and child" against swine flu it was probably an unfortunate term to use. Hopefully it would not come back to haunt them.

So far the medical community was guardedly optimistic. Thirty thousand volunteers in Dawn's trials would either receive two shots of placebos or two doses of vaccine. Then they would have to wait to see if they would be exposed to Covid-19 and if they did would their immune systems be able to fight off the illness. Dawn testified that it was a painless procedure, over before she realized she had even been given the shot.

Hopefully one or more of the vaccines would work and the world would also recover before it even realized what had happened.

CHAPTER 44
Russian Roulette
August 11, 2020

On August 11 Vladimir Putin announced that Russia had registered the world's first Coronavirus vaccine. Standing beside him was his minster of health Mikael Murashko who reported that all the volunteers who had been participated in the vaccine's Phase 1 and Phase 2 trials had developed high titers of antibodies and nobody had suffered any complications.

Therefore the Phase 3 trials could begin on the 12th and would involve 2,000 people from Russia, The United Emirates, Saudi Arabia, Brazil and Mexico. Then, full-scale production would begin in September.

With this announcement Putin hoped to renew some of the prestige Russia enjoyed when Ilya Metchnikoff had discovered the fundamentals of immunology and the Soviet Union had developed a new smallpox vaccine and initiated the program to wipe the scourge off the face of the planet.

It sounded almost too good to be true and American analysts were quick to say so.

J. Stephen Morrison from the Center for Strategic and International Studies thought that Putin was trying to just score a domestic win after bungling Russia's outbreak and failing to revise the economy.

"Let me be very blunt here; Putin needs a win, he needs a domestic win. He mismanaged the outbreak within his own territory. He's lost the public trust and confidence in his efforts; his economy is on its back. He can't deliver on any of his big infrastructure projects that he promised in the electoral campaign." *

Scientists voiced similar concerns. The Gamaleye Research Institute had registered the Phase 1 and Phase 2 tests for their vaccine but it had only tested it on 38 volunteers and it hadn't even published its data. This raised red flags for Natalie Deane of the University of Florida.

"The timing of the Russian announcement makes it highly unlikely that they have sufficient data about the efficacy of the product."
Cornell medical School's John Moore agreed. "Putin doesn't have a vaccine he just has a political statement."

"The lesson that the U.S. government should draw from Mr. Putin's announcement is clear," said Nichole Lurie from the US Department of Health. "This is exactly the situation that American should try to avoid." But John Moore had the last word, "This is all beyond stupid." *

An American Solution

Meanwhile, Michael Mina, an epidemiologist from Harvard's Chan School of Public Health used a digital press conference to announce what could become a distinctly American solution to its runaway pandemic problem. It was a small cheap do it yourself paper-based Covid testing device similar to an at home pregnancy test.

Two small start-up biotech companies and Minnesota's multinational 3M Corporation already had prototypes of the devices and could put hundreds of millions of the kits into the public's hands within weeks of the hoped for FDA approval.

If everyone who used the device and tested positive stayed home, it could stop the disease in its tracks. It would be like having an instant vaccine, an overnight way to create herd immunity artificially.

The device had been developed at the joint Harvard MIT Wyss Institute that had produced a similar paper-based test for tick borne diseases.

The only kicking point for the FDA was that the Wyss test was less accurate than lab based tests. But Mina estimated that the lab-based were done so infrequently and took so long to produce that they only caught 3% of the cases early enough to stop people from spreading the disease, but were highly efficacious at telling a person what he had just died from.

Mina closed his telephone press conference with the admonition that these devices were our best hope; because we didn't have anything else we could do tomorrow, other than to keep the schools closed and shut down the economy once again. **

* CNBC 8/12/2020. Russian Vaccine at High Risk of Backfiring.
**Harvard Gazette 8/12/2020.

CHAPTER 45
The Worst Case Scenario
Ipswich, Massachusetts
August 13, 2020

By mid-August it looked like the United States had squandered its opportunity to tampen down the transmission of Covid-19 so schools could open in the fall.

Had we put our summer recreation before our children's safety? The few schools that had tried to open early had to be shut down again after teachers and students contracted the disease on the first week of class. But states were continuing to hold rodeos and massive motorcycle events as if the problem had been solved by a few weeks of shutdowns.

Over 730,000 people had already died and the world's death toll would probably surpass a million before any vaccines could start making any impact. More US citizens were getting Covid every week than had contracted the disease in Great Britain throughout the entire course of the pandemic. We had an astonishing 5 million cases.

But what if the very worst happens, what if the vaccines don't work or are only partially successful? What if people refuse to get the shots and continue to re-infect others?

Who can guarantee that the number wont keep climbing close to the 50 million people who died in the 1918 Spanish flu? That would still be less than the 70 million people born every year on average. Plus, countries might gain a surfeit of Covid babies nine months after their lockdowns had gone into effect.

But diseases can do much more than just slow growth; they can cause a population to precipitously crash.

One of the classic examples in the animal kingdom are gypsy moth caterpillars. Their numbers grow exponentially for several years until they contract a polyhedrosis virus that liquefies their innards.

The virus then releases an enzyme so the caterpillars' exoskeletons split apart and the caterpillars melt into the leafy substrate leaving only a smear of virions ready to infect the next hatch of gypsy moths. After a year or so of such repeated infections the population of gypsy moths crashes and it can then take decades for the moths to recover.

The same thing can happen to larger animals, like the rabbits that were introduced to Australia where they had no natural enemies and North Atlantic seals whose numbers surged after conservation officials removed bounties that had kept their numbers in check to prevent the spread of cod worms.

Plus populations of deer have exploded because we killed off their wolf and mountain lions predators and let farmlands revert to forestlands.

The Australian rabbits were initially killed off by a myxoma virus though they eventually developed an immunity to the introduced pathogen.

A wasting disease is attacking western deer and North Atlantic seals have succumbed to a flulike porcine distemper. The collapse of the seal population lead to a plethora of Great White Sharks who seem to be developing a taste for human flesh in the seals' absence.

The same thing can also happen to humans. The Plague wiped out 30 to 50% of the population of Europe. If you can call anything about the Black Death a silver lining, for years people lived longer and were healthier after the pandemic than before. Apparently the pandemic had culled out older people and those with co-morbidities, certainly not the sort of thing you want to encourage.

But now, humans comprise a hundred times more biomass than any other large animal that ever lived on our planet. And other than an occasional snakebite, shark attack, or marauding lion we have little to fear from large predators. In the words of David Quamann, the author of Spillover, we are large, long-lived and grotesquely abundant.

We and our livestock have exceeded the natural carrying capacity of the earth several times over. We have created the environment where a virus either spilled out of a lab or out of a reservoir organism and we don't really know if all our lockdowns, quarantines and isolations or even if our vaccines will be enough to stop this rapidly mutating scourge for several years.

But we do know that we are living through a watershed event. As the editor of the medical journal Lancet Richard Horton wrote in his book, The Covid-19

Catastrophe, "Maybe the truth is that life will never fully return to normal, perhaps Coved -19 represents an impenetrable boundary between one moment in our lives and another. We can never go back...."

Postscript
Three Arthropods
Ipswich 2020

As a kid growing up on Cape Cod, I was obsessed with three arthropods; horseshoe crabs, pill bugs and dare I say it, ticks.

In the mornings I would lie in bed savoring the sounds of songbirds swelling to greet the newborn day. As they reached their crescendo the first rays of sunshine would stream through my window suffusing my knotty pine walls with its warm yellow glow.

I would slip into a bathing suit and try to tiptoe down our creaky back stairs without waking my recalcitrant family. But in my eagerness to get outside I would inevitably burst through the screen door, which would slam shut behind me.

"Go back to sleep!" My older sister would bellow from her nearby bedroom.

Outside the sun would shine on a pair of tiny young rabbits timidly nibbling the dew-drenched grass. They would pause to stare up with deep dark eyes. Then with a twitch, and a flicker they would bound beneath the fragrant thicket of bayberry already abuzz with honeybees.

I would walk quietly to the row of tomato plants that grew beneath my sister's window. They flourished there, bathed in the first light of sun, yet still protected from onshore winds.

The vines would be beaded with drops of shimmering dew, caught in the tiny hairs of their succulent leaves. As I brushed through the tomatoes I would be enveloped in the smell of their leaves. It was the tangy odor of the earth, the sun—of summer itself.

I would pluck a tomato and bite into its sun-warmed flesh. Cool juices would explode in my mouth and dribble down my bare chest. I was eating the earth, drinking the sun, inhaling the universe, exhaling the raw materials of life itself.

Then I would sit lizard-like in the sun, entranced by the Paleozoic world at my feet. Pill bugs were rolling themselves up into perfectly round spheres, as their ancestors had been doing for hundreds of millions of years.

I could not be more in the present, more childlike, more attuned to the rhythms of our ancient ever-changing universe. I was but a tiny ephemeral mote in the cosmic intricacy of life itself.

"Bill-ly!"

My reverie would be instantly broken. My family had arrived en masse to partake of their indoor, to my mind far more pedestrian meal. They would argue and plan, bicker and gulp down great spoonfuls of politically incorrect consumables.

I never let on that I had just been dining with the gods, been drunk on the essence of the universe, lost in the world of ancient arthropods.

The middle of my day would be devoted to studying the behavior of horseshoe crabs and I'm embarrassed to say it also involved spearing them with an oar because the state considered the gentle arthropods shellfish predators. I'm afraid I can still feel the crunch of their shells as we impaled them on our makeshift spears.

The end of my day would be taken up with de-ticking our dogs. The entire family would sits on our splintery wooden porch combing through the hair of our two Labrador retrievers. When we found a nice big fat dog tick we would pick it off with tweezers and plop it into a jar of kerosene teeming with the summer's harvest of hundreds of still living arachnids.

Occasionally I would use my beloved hand lens to focus a pinpoint of sunlight on the back of an unfortunate tick until it browned and sizzled emitting an odor if not as civilized as a madeleine just as redolent of things long past.

My transgressions were mitigated, somewhat, by my scientific interest in the creatures. I would scrutinize them under my microscope and illustrate their most salient features.

I even saved one particularly bloated tick in a bottle. She eventually converted all her dog's blood into thousands of perfectly round eggs, which I sent off to the Boston Museum of Science along with several mounted greenhead flies.

But the curious thing was, although we removed thousands of dog ticks and were bitten by quite a few; we never found a deer tick and never got sick.

Today, we can no longer get together in our family home because of Covid-19, my granddaughter can no longer play in the grass because Lyme disease, and the only way we are going to be able to rid our planet of these scourges is to use the blood of horseshoe crabs to check that our vaccines and antibody tests are pyrogen free.

On 9/11 I was finishing a book about the horseshoe crab test and found myself asking the question why I was spending my time writing about horseshoe crabs when thousands of people had just died, anthrax was being sent through the mail and United States troops were being vaccinated against biological weapons they expected to encounter in Iraq.

John Steinbeck asked the same question during World War II when bombs were being dropped on Europe and he was writing about collecting starfish in the tidepools of the Sea of Cortez. He decided that, "All of it was important or none of it was important at all."

But both of us had been asking the wrong question. It made little difference that we had been writing about starfish and horseshoe crabs when thousands of humans had died. Each organism was part of that intricately complex unity of life I had felt so intuitively as a kid.

The real question raised by the Covid Pandemic was not how it originated and propagated throughout the world, but how the unity of life is so intricately complex and the indispensible to our existence and future.

Here was a disease of humans caused by a virus evolved in a complex mix of bats, ticks, snakes, pangolins plus a laboratory ferret or other, as yet unidentified organism, that can only be thwarted with the aid of a spiderlike arachnid who first crawled out of te ocean 450 million years ago.

Life itself had probably started from such a virus-like piece of RNA that existed in the twilight world between the living and the dead. It had evolved through the archaebacteria into nucleated organisms like horseshoe crabs, ticks and spiders and on to the bats, fish snakes and pangolins concentrated in the wet seafood market in Wuhan, when they should have been scattered throughout the world. .

We have discovered the power of DNA and found it is just as dangerous to splice, dice and manipulate DNA, as it is to split atoms, just as dangerous to make biological weapons as to make nuclear bombs.

We need to unmask and fear uncontrollable biological weapons as much if not more so than the atomic bombs. One can destroy our civilizations the other our very existence, in this fragile experiment we call life.

We should take equal care when we deforest our continents to raise palm oil, soy and beef cattle, and when we eat bush meat so close to the zoonotic filled wilderness areas. We have inherited this vast wonderful interconnected biosphere. It will survive our depredations, but we continue to splice dice and tweak it at our own species peril.

NOTES

CHAPTER ONE
The Death of a Nation Patuxet, Massachusetts
November 1617

This chapter is based on information gleaned from current and historic sources as reported in the cited material. The original sources include archaeological evidence, Native American studies and the writings Governor Bradford and the early explorers of New England. While the details are accurate as reported the conversations and characters are fictitious.

1) Descriptions of Wampanoag village life come from research reported in Paul Schneider, The Enduring Shore, Henry Holt Co. NY 2000. P26, 32-33.

2) Patuxet was the Wampanoag name for the village that predated the Plymouth Plantation. When the Pilgrims arrived they found fields already cleared for agriculture but the inhabitants missing. They had all died from what is believed to have been smallpox introduced by European explores and fishermen prior to settlement.

3) Many of the native American names and descriptions of Wampanoag life come from Earl Mill's and Alicja Mann's book, Son of Mashpee, Recollections of Chief Flying Cloud, a Wampanoag, Word Studio, Falmouth, MA 1996.

4) Descriptions of the artifacts in Attaquin's dugout come from an archeological site recently discovered in Quincy Massachusetts. The artifacts indicate that the dugout canoe was fully loaded to spear swordfish in offshore waters or to flense and recover whale meat from beached pilot whales. Governor Bradford of the Plymouth Plantation writes about treaties made between sachems of Native American villages to allocate this valuable resource that often meant the difference between a winter of plenty or a winter of scarcity.

5) Earl Mills, former tribal chief of the Wampanoags, told me the story that after every Thanksgiving his relatives would sit around the table and joke that if their ancestors had not shown the white men how to make such good food they would have probably gone home. I like to think that this story was handed down through several generations.

6) The description of this Wampanoag ball game similar to Lacrosse was described in The Enduring Shore, Paul Schneider, Henry Holt Co. NY, 2000, p67.

7) Symptoms of Smallpox from Elizabeth A. Penn's excellent piece of scholarship, Pox Americana, Farrar, Straus and Giroux, NY ply 9.

8) Paul Schneider, The Enduring Shore, Henry Holt Co. 2000 NY p 70

9) " ... for many weeks together" Ibid p69

10) ibid p117

CHAPTER TWO
Washington's Gamble

This chapter is based on descriptions of the variolation of troops during the American Revolution as described in Elizabeth's Penn's fascinating book, Pox Americana.

I) Such tactics were condoned by even such honorable men as Lord Jeffrey Amherst who gave his name to Amherst College. Elizabeth Fenn, Pox Americana, Farrar Strauss & Giroux, New York, 2001, p88

2) ibid description of smallpox quarantines during the occupation of Boston. P49-5l

3) Description of lord Dunsmore's Ethiopian regiment ibid p55-57.

4) Cotton Mather's description of the first use of variolation in America in Elizabeth A Fenn, Pox Americana, Farrar, Straus and Giroux, NY p32

5) John Adams' description of his variolation. Ibid p93-98.

6) Ibid description of George Washington's trip to Barbados where he caught smallpox. Pl3-15

7) Description of variolation in Ethan Allen's troops. Ibid p33-35

8) Lafayette, ibid p

9) " ... Hessian prisoners of war." Ibid p98

10) Ibid p99

11) description of self variolation ibid p33

CHAPTER THREE
Four Scientists
1796-1928

Much of this chapter is based on descriptions drawn of Paul de Kruifs classic book, Microbe Hunters. More than a few scientific careers were whetted by reading this stirring narrative.

1) M.D. Anderson, Through the Microscope; Man looks at an Unseen World. Natural History Press, NJ, 1965, p84.

2) Ibid p85.

3) The Jenner Museum websitewww.fivevalleys.demon.co.uk/jenner.html

4) Description of Robert Koch's work on anthrax in Paul de Kruif, Microbe Hunters, Harcourt Brace Orlando, Florida, 1926, pl0l-139.

5) Ibid description of anthrax in Wollstein, pl05.

6) Description of Koch's method of injection. Ibid pl09

7) Description of Koch's hanging drop technique. Ibid p. 14

8) Description of anthrax etiology. Ibid p117

9) Description of Louis Pasteur's work on anthrax. Ibid pl40-l 77.

10) "They are gay. They are eating." Ibid pl48

11) "I have demonstrated ... " at the Academy of Medicine, Ibid ply 50.

12) Description of the events at Pouilly-le-fort. Ibid pl58

13) "The experiment... is an unprecedented success." De Blowitz quote in The London Times.

Ibid pl62

14) "Such goings on... " Koch quote. Ibid pl 62.

15) "Dr. Thullier, of the French Commission ... " ibid pl36

16) "...but they are laurel, such as are given to the brave." Ibid pl36

17) Description of Fleming's Discovery of penicillin in A Science Odyssey; People and Discoveries, Alexander Fleming. 1998, Pbs.org/wgbh.

CHAPTER FOUR
Two Planes
The Summer of 1942

I) BBC News Online, July25, 2001, news.bbc.co.uk.

2) Eileen Choffness, Germs on the Loose, Bulletin of the Atomic Scientists, March-April 200 I.

3) BBC News online July 25, 2001

4) ibid

5) Description of the attack on Congshan. Tom Mangold and Jeff Goldberg, Plague Wars; The Terrifying Reality of Biological Warfare, St Martins Griffin NY, I999, p22-23.

6) "It's a war, you do what you have to do." Ibid p20

7) Description of Ping Fan. Ibid pl 7-18.

8) Description of porcelain bomb. Ibid 24-25.

CHAPTER EIGHT
The Test
The Dugway Proving Grounds, Utah
July 12, 1955

This chapter uses several sources to tell the story of Seven Day Adventist's who volunteered to be exposed to biological weapons during Operation White Coat.

I) Aaron Zitner, Taking a Germ Bullet, LA Times, November 26, 2001

2) Judith Miller Stephen Engelberg William Broad, Germs; Biological Weapons and America's Secret War, Simon and Schuster, NY, 2001, p42

3) "terrible and inhumane" ibid p38

4) History of Camp Detrick. Ibid p38-39

5) Description of Theodore Rosebury's book. Ibid p41

6) Description of the Pine Bluff facility. Ibid p50

7) Possible use of biological weapons by the Soviet Union in Afghanistan. Ibid p77

8) "One great difficulty." Ibid p55

9) "Let me think about. I'll write you some papers." Ibid p62 I 0) Ibid p63

CHAPTER TEN
The Research Paper Stalingrad, 1942

Chapter Five is based on interviews with Ken Alibek and his invaluable book, Biohazard. I assumed that I would have to write several letters and wade through mounds ofred tape to interview a man who used to travel with two KGB bodyguards and had been in charge of the Soviet Union's Biological weapons program. But one day I just called and, to my astonishment, who should answer but Ken Alibek. From then on I made almost weekly calls to ask everything from what the weather would be like in Moscow in October to how to spell a particular Russian expression. We became "Ken" and "Bill" as I grew to appreciate an extraordinarily open and brave scientist who rose to the top of the Soviet Union's Biological Weapon's Program and now serves the world and the United States through his knowledge of biological warfare and ways to prevent it.

I) Ken Alibek with Stephen Handelman, Biohazard; The Chilling True Story of the Largest Covert Biological Weapons Program in the World - Told from Inside by the Man Who Ran it. Delta, Random House, NY, 1999, p28-29.

2) Conversation between Ken Alibek and Colonel Aksyonenko. Ibid p30-3 I.

3) Boris Pasternak, Dr. Zhivago, Pantheon Press, NY 1958, p365-366.

4) Description of the smallpox program at Zagorsk in Alibek, Biohazard, p112.

5) Description of Soviet smallpox program. Ibid p298

6) Description of Soviet biology under the thumb ofTrofim Lysenko. Ibid p39-40

7) Description of Ovchinnikov's rescue of Soviet biology by linking it to biological; warfare. Ibid p40-4 I.

CHAPTER ELEVEN
Tightening the Noose
The Smallpox Eradication Program
1958-1978

I) William Sargent, The Year of the Crab; Marine Animals in Modern Medicine, WW N01ton Co. NY 1995, p 62?

2) Dimitri Ivanofski, Scientist Notebook, 1996www.New—scientist.com

3) Ken Alibek with Stephen Handelman, Biohazard, Delta-Random House NY, 1999, pl 11

4) Laurie Garrett, The Coming Plague, Penguin Books, NY, 1994, p41

5) Ibid p42.

6) Breakthrough Medicine, August 6, 1997, www.omnimag.com/archives p3

7) Laurie Garrett, The Coming Plague, Penguin Books, NY 1994 p45

8) Ibid p45 (?)

9) Breakthrough Medicine, August 6, 1997, www.omnimag.com/archives, p3-4

10) Laurie Garrett, the Coming Plague, penguin Books, NY 1994, p45

CHAPTER TWELVE
"To Innoculate Every Man Woman and Child." The Swine Flu Non-Epidemic 1976

Much has been written about the 1976 Swine Flu crisis. Readers who wish to pursue this topic further should start with Laurie Garrett's comprehensive and readable account in her excellent 1994 book, The Coming Plague; Newly Emerging Diseases in a World out of Balance.

1) "60 Minutes" Sunday Nov 4, 1979

2) Garrett, Laurie The Coming Plague, Penguin Books (1994) p154

3) Ibid

4) Fineberg, Harvey and Neustadt, Richard The Swine Flu Affair; Decision-Making on a Slippery Slope. Pl63

5) President Ford Press Conference, quoted in "The Coming Plague" pl64

6) Silverstein, Arthur quoted from "The Coming Plague" p164

7) Garrett, Laurie The Coming Plague (1994) p167

8) Ibid. p 167

9) Ibid. p 172

10) "60 Minutes" November 4, 1979

11) Crab Wars; A Tale ofHorsehoe Crabs, Bioterrorism and Human Health. University Press of New England (2002) p 51

12) The Coming Plague p617

CHAPTER SIXTEEN
Sverdlosk 1979

Many books and articles have been written about Sverdlosk; they all disagree slightly on some of the details of the incident. Anthrax: The investigation of a deadly outbreak provided excel lent descriptions of the city and their investigation. Germs and The Plague Wars provided the diplomatic and political background and Biohazard provided the Soviet perspective. I have drawn from these and interviews with Ken Alibek to try to tell a comprehensive story of what was happening in the Soviet Union and the United States during this period.

1) Piller and Yamamato, Germ Wars p134

2) Miller J, Engelberg S, Broad W, Germs; Biological Weapons and America's Secret War, Simon and Schuster, NY, 2001 p84

3) Ibid p 80.

4) Guillemin Jeanne; Anthrax; University of California Press 2001 p37

5) Alibek Ken, Biohazard Delta Book Random House 2000 p73

6) Guillemin Jeanne; Anthrax; University of california Press, 2001 p4

7) Ibid Alibek, p78

8) Ibid p73

9) Young John, Collier R. John; Scientific American March 2002 p3 HMLT

10) Mangold Tom, Goldberg Jeff The Plague Wars St Martin's Griffin 2001 p69-70

11) Ibid p 70

12) Ibid p 72

13) Alibek p84

14) R. Weiss Science News, Vol 133 April 88 Meselson, p9,51

15) Mangold p80

16) Smith R. Jefferey, Hilts Phillip J. Soviets Deny Lab Caused Anthrax"
Washington Post April 13 1988 Al

17) Alibek p 86

18) Mangold p80 I9) Alibek p 79

CHAPTER SEVENTEEN
The Phone Call
Biopreparat October
1989

Most of the events described in this chapter came from a series of e-mails and
telephone interviews with Ken Alibek, others were quoted from Biohazard his
fascinating book co-authored with Stephen Handelman.

I) Ken Alibek telephone interview January 4, 2002

2) Alibek Ken and Handelman Stephen Biohazard Dell Publishing, New York, 1999
p.137.

3) Ibid p.81

4) Ibid p.88

5) Ibid p.117-118

6) Mangold Tom, Goldberg Jeff, Plague Wars, Macmillan Publishers, London, 1999
p.96

7) Alibek, p.10

8) Ibid pl37

9) Ibidp.141

10) Ibid p.150

11) Ibid p. 182-183

12) Ibid p.195-196

13) Mangold Tom, Goldberg Jeff, Plague Wars Macmillan Publishers, London, 1999 p.96

14) Croddy Eric, Chemical and Biological Warfare, Copernicus Books, New York, 2002 p.208

15) Alibek Ken and Handelman Stephen Biohazard, Delta Random House, New York, p. 199- 200

16) Ibid p. 203-204

17) Ibid p.204

18) Ibid p.204

CHAPTER EIGHTEEN
Defection October 1992

Most of the events and conversations related in this chapter came from a series of telephone interviews with Ken Alibek in 2002 and from his book Biohazard; The Chilling True Story of the Largest Covert Biological Weapons Program in the World, written with Stephen Handelman.

I) Ken Alibek with Stephen Handelman, Biohazard, Random House, New York, 1999 p228

2) Tom Mangold and Jeff Goldberg, Plague Wars; The Terrifying Reality of Biological Warfare, St. Martin's Press, New York, 1999 pl59

3) Ibid pl47

4) Biohazard, pl97

5)Ibid, p236

6)Ibid, p229

7) Ibid, p238

8) Ibid, p240

9) Ibid, p250-25 I

10) Plague Wars pl95

11) Ibid p!98

12) Ibid pl98

13) Ibid p209

CHAPTER NINETEEN

6) Ibid from notes made by one of the participants as reported in Germs. p237

7) Ibid p238

8) Ibid p219

9) Ibid p240

10) Ibid p239

11) Ibid p241

12) Ibid p241

13) Ibid p241

14) Ibid p242

15) Tom Monath interview June 2002

CHAPTER TWENTY
The Bomblet Washington, 1999

I) Judith Miller, Stephen Engelberg, William Broad; Germs; Biological Weapons and America's Secret War, Simon and Schuster, New York 2001 p 293.

2) Ken Alibek interview July 6, 2002

3) Judith Miller, Stephen Engelberg, William Broad, germs; Biological Weapons and America's Secret War, Simon and Schuster, New York, 2001, p288

4) Ibid p.293

5) Ibid p.297

6) Ibid p.298

7) Ibid p.298

8) Ibid p.297

9) Ibid p 309

10) Ibidp310

CHAPTER TWENTY- THREE
Outbreak at Aralsk The Monterey Report
June 14, 2002

This chapter is based on a report published by the Monterey Institute as reported in the New York Times in June 2002. It was supplemented with conversations with Ken Alibek and by the author's experience working as a research assistant aboard Soviet Fisheries vessels in the 1970's.

1) William Broad and Judith Miller, The New York Times, June 15, 2002

2) The Monterey Report did not specify which animal if any were used in this particular test. In an interview on June 26, 2002, Ken Alibek told the author that Soviet scientists had perfected the Hamadryas baboon as an animal model for smallpox and that they were the most likely animals used in this test. He has witnessed several similar tests and provided details about tehering a hundred animals to poles the night before a test. This was also not specified in the Monterey Report.

3) This conversation is based on similar conversations the author had as a research assistant aboard two Soviet Fisheries Vessels, the RN Blesk and the RN Belegorsk.

4) Ken Alibek interview June 26, 2002.

5) Ibid

6) Ibid

7) Ibid

8) Lawrence Altman, The New York Times, June 21" 2002.

9) Ken Alibek interview June 26th 2002

I 0) Editorial, The New York Times, June 22, 2002.

CHAPTER TWENTY-SIX
Eckard Wimmer; The Polio Maker
July I I, 2002

I) Robin Finn, New York Times, July 19,2002

2) Ibid

3) Interview with Dr. Wimmer's secretary July 21, 2002.

4) Andrea Pollack, New York Times, July 12, 2002

5) Judith Miller, Stephen Engelberg and William Broad, Germs; Biological
Weapons and America's Secret War; Simon and Schuster, NY, 2001 p308

6) In The Cobra Event published by Ballantine Books in 1997, Richard Preston
had his protagonist use the "wiping the slate clean" argument as his justification for
attacking New York City with biological weapons.